A CULTURAL HISTORY OF MATHEMATICS

VOLUME 6

A Cultural History of Mathematics
General Editors: David E. Rowe and Joseph W. Dauben

Volume 1
A Cultural History of Mathematics in Antiquity
Edited by Michael N. Fried

Volume 2
A Cultural History of Mathematics in the Medieval Age
Edited by Joseph W. Dauben, Clemency Montelle, and Kim Plofker

Volume 3
A Cultural History of Mathematics in the Early Modern Age
Edited by Jeanne Peiffer and Volker R. Remmert

Volume 4
A Cultural History of Mathematics in the Eighteenth Century
Edited by Maarten Bullynck

Volume 5
A Cultural History of Mathematics in the Nineteenth Century
Edited by Tom Archibald and David E. Rowe

Volume 6
A Cultural History of Mathematics in the Modern Age
Edited by Tom Archibald and David E. Rowe

A CULTURAL HISTORY OF MATHEMATICS

IN THE MODERN AGE

VOLUME 6

Edited by Tom Archibald and David E. Rowe

BLOOMSBURY ACADEMIC
LONDON • NEW YORK • OXFORD • NEW DELHI • SYDNEY

BLOOMSBURY ACADEMIC
Bloomsbury Publishing Plc
50 Bedford Square, London, WC1B 3DP, UK
1385 Broadway, New York, NY 10018, USA
29 Earlsfort Terrace, Dublin 2, Ireland

BLOOMSBURY, BLOOMSBURY ACADEMIC and the Diana logo are trademarks of Bloomsbury Publishing Plc

First published in Great Britain 2024

Series design by Raven Design
Cover image: A girl solves a mathematical equation at a school created for children not previously permitted to attend regular schools. Kabul, Afghanistan © Majority World/ Universal Images Group via Getty Images

A catalogue record for this book is available from the British Library.

A catalog record for this book is available from the Library of Congress.

ISBN: HB: 978-1-3500-6307-5
ePUB: 978-1-3502-9495-0
ePDF: 978-1-3502-9494-3
Set: 978-1-3500-6312-9

Series: The Cultural Histories Series

Typeset by RefineCatch Limited, Bungay, Suffolk
Printed and bound in Great Britain

CONTENTS

ILLUSTRATIONS

Introduction

Chapter 1

Chapter 2

Chapter 5

Chapter 6

Chapter 7

CONTRIBUTORS

Tom Archibald took his PhD in History and Philosophy of Science at the University of Toronto, Canada, and is currently Professor of Mathematics at Simon Fraser University, Burnaby, Canada. His research concerns the history of mathematical analysis and its applications, and the relationship between mathematics and society more broadly. Until recently, he was co-editor-in-chief of *Historia Mathematica* and a member of the Executive Committee of the International Commission on the History of Mathematics. He now serves on the editorial board of the *Revue d'histoire des mathématiques*. Tom co-authored with Della Dumbaugh and Deborah Kent, "A Mobilized Community: Mathematicians in the United States during the First World War," in David Aubin and Catherine Goldstein eds., *The War of Guns and Mathematics* (2014); and with Rossana Tazzioli, "Integral Equations between Theory and Practice: The Cases of Italy and France to 1920" *Archive for History of Exact Sciences* (2014).

Roger Cooke received his PhD in Mathematics from Princeton University, New Jersey, USA, in 1966, afterward joining the faculty at the University of Vermont, USA. In 1980, he began doing research on the history of mathematics, which led to the publication of his first book, *The Mathematics of Sonya Kovalevskaya* (1984). His most recent publications are *The Case of Academician Nikolai Nikolaevich Luzin* (a translation from Russian, 2016) and *It's About Time: Mathematical Reflections on the Theory of Relativity* (2017), both published by the American Mathematical Society.

Leo Corry is the Bert and Barbara Cohn Professor of History and Philosophy of Exact Sciences at Tel Aviv University, Israel. His main research interests cover the history of mathematics, physics, and computation, beginning in the late nineteenth century, as well as the Euclidean tradition in the Medieval and Early Modern periods. His latest publications include *Chaim L. Pekeris and the Art of Applying Mathematics with WEIZAC, 1955–1963* (2022), and *British Versions of Book II of Euclid's Elements: Geometry, Arithmetic, Algebra, 1550–1750* (2022).

Liesbeth De Mol is an art historian and philosopher working on the history and epistemology of computing and programming. She co-founded the DHST/DLMPST commission for the history and philosophy of computing and is the PI of the ANR research project PROGRAMme on the history and philosophy of computer programs. She is a permanent researcher with the Centre national de la recherche scientifique (CNRS), Paris, and is affiliated with the University of Lille, France. Liesbeth is the author of "Logic, Programming, and Computer Science: Local Perspectives," *IEEE Annals of the History of Computing* (2021); and co-authored with Maarten Bullynck, "What's in a Name? Origins, Transpositions and Transformations of the Triptych Algorithm–Code–Program," in Janet Abbate and Stephanie Dick eds., *Abstractions and Embodiments: New Histories of Computing and Society* (2022).

Nina Engelhardt is a scholar of English literature at the University of Stuttgart, Germany. She received her PhD from the University of Edinburgh, UK, and is author of the monograph *Modernism, Fiction and Mathematics* (2018) as well as of a number of articles in the area of literature and mathematics studies. Together with Robert Tubbs and Alice Jenkins, she is co-editor of *The Palgrave Handbook of Literature and Mathematics* (2020).

José Ferreirós is Professor of Logic and Philosophy of Science at the Universidad de Sevilla, Spain. His interests are in the history of mathematics and the sciences in the nineteenth and twentieth centuries, and the philosophy of mathematics. Among his publications are *Mathematical Knowledge and the Interplay of Practices* (2016) and the collective volume *The Architecture of Modern Mathematics* (2006). He is a founding member of the Association for Philosophy of Mathematical Practice and serves on the editorial board of *Annals of Mathematics and Philosophy*.

Christopher J. Phillips is Associate Professor of History at Carnegie Mellon University, Pittsburgh, Pennsylvania, USA, where his principle research interests are in the spread of mathematical methods into new domains. He received his PhD in History of Science from Harvard University, Cambridge, Massachusetts, USA. His books include *The New Math: A Political History* (2014) and *Scouting and Scoring: How We Know What We Know About Baseball* (2019). He is currently finishing a monograph entitled *Number Doctors: The Emergence of Biostatistics and the Reformation of Modern Medicine*.

David E. Rowe is Professor Emeritus of History of Mathematics and Natural Sciences at Mainz University, Germany. His principal research interests concern developments in Western mathematics and physics during the period 1800 to 1950, with special focus on Göttingen University, Germany. Since 2011, he and Klaus Volkert have co-edited the series *Mathematik im Kontext*. He has written or co-edited sixteen books, most recently *Emmy Noether: Mathematician Extraordinaire* (2021).

Tilman Sauer studied Physics, Mathematics, and Philosophy at the Free University of Berlin, Germany. After postdoctoral positions at the Max Planck Institute for History of Science in Berlin and at the Universities of Göttingen and Bern, Germany, he was Research Associate at the California Institute of Technology, USA, co-editing the Collected Papers of Albert Einstein. In 2008, he obtained his habilitation at the University of Bern. Since 2015, he teaches History of Mathematics and Natural Sciences at Johannes Gutenberg University in Mainz, Germany. Among his publications are: "Einstein's Unified Field Theory Program," in Michel Janssen and Christoph Lehner eds., *The Cambridge Companion to Einstein* (2021); and as co-editor with Tilman Sauer and Ulrich Majer eds., *David Hilbert's Lectures on the Foundations of Physics, 1915–1927: Relativity, Quantum Theory, and Epistemology* (2009).

Marta Jordi Taltavull studied physics in Barcelona before beginning doctoral studies at the Max Planck Institute for History of Science in Berlin, Germany. After taking her PhD at Humboldt University, Germany, she held a postdoctoral position at Mainz University, Germany, from 2017 to 2020 before her appointment as Scientific Coordinator of the Menorcan Institute of Studies, Mao, Menorca, Spain. Her primary research interests concern the history of quantum physics and its relationship to the history of optical phenomena. Among her publications are "Sorting Things Out: Drude and the Foundations of Classical Optics," in Massimiliano Badino and Jaume Navarro eds., *Research and Pedagogy: A History of Quantum Physics through Its Textbooks* (2013); and "Transmitting Knowledge across Divides: Optical Dispersion between Classical and Quantum Physics," *Historical Studies in the Natural Sciences* (2016).

SERIES PREFACE

Culture, history, and mathematics are words that seldom appear together. Yet, mathematical activities have played a pervasive role in diverse cultures throughout human history, as described in the present six volumes. These illustrate how individuals and societies have engaged with various manifestations of mathematical culture from antiquity to the present. Based on recent and ongoing scholarship, they provide a global overview of a highly diverse range of developments, though certainly nothing like a comprehensive picture. Editors and authors had to make difficult choices about which topics and cultures to include, and in doing so, they have sought to strike a balance between characteristic features of a period and particular individuals and events that had a deep influence or longer lasting impact.

Each volume addresses seven general themes in as many chapters. The first three chapters —Everyday Numeracy, Practice and Profession, and Inventing Mathematics—highlight the roles of those who used, taught, or created mathematics as it evolved over time. These chapters explore mathematics per se, whereas the three that follow—Mathematics and Worldviews, Describing and Understanding the World, Mathematics and Technological Change—consider the larger impact of mathematics on other dimensions of human culture. The final chapter—Representing Mathematics—is devoted to a broader understanding of mathematical ideas, activities, or artifacts in different cultures throughout the world. The periods covered are:

Volume 1: A Cultural History of Mathematics in Antiquity
(3000 BCE–500 CE)
Volume 2: A Cultural History of Mathematics in the Medieval Age
(500–1400)

Volume 3: A Cultural History of Mathematics in the Early Modern Age (1400–1687)
Volume 4: A Cultural History of Mathematics in the Eighteenth Century (1687–1800)
Volume 5: A Cultural History of Mathematics in the Nineteenth Century (1800–1914)
Volume 6: A Cultural History of Mathematics in the Modern Age (1914–2020)

General Editors: David E. Rowe and Joseph W. Dauben

Introduction

TOM ARCHIBALD AND DAVID E. ROWE

The period covered by this volume witnessed enormous political, social and economic turmoil along with extraordinary technical developments. The world population in 1920 reached nearly 2 billion people; today, more than four times that number inhabit the planet. Yet, the learned world—and especially those with specialized scientific and mathematical training—has grown even more dramatically. Science and technology now interact reflexively to transform each other, a process that gained in rapidity with the passage into the "information age," as mathematics has become directly involved with technological change. So pervasive is technology that it implicates nearly everyone. Hence, the world we live in today is full of the often hidden mathematics that underlies those more visible and familiar technologies.

The century discussed in this volume falls into three periods: 1914–45, beginning and ending with the two World Wars; 1945–90, dominated by the Cold War between East and West and ending with the collapse of the Soviet Union; and 1990 to the present, the moment of the COVID-19 pandemic and general acknowledgment of human-generated climate change. Each of these eras possesses distinctive features as regards mathematics and its relation to society. The prominent role of the United States, both economically and culturally, serves as a guiding thread throughout much of this period.

A key shift for mathematics in society has been its role in warfare and defense. While still relatively rudimentary at the time of the First World War, the importance of sophisticated mathematics became crucial for the development of the atomic bomb along with the aircraft and rocketry of the Cold War. The omnipresence of mathematical technologies in computing and security established itself gradually in the period after the Second World War. These military technologies likewise spilled over into civil society: the transmission of digital signals, the encryption of banking and medical information, the development of cellular phones, transportation systems, drug testing, and infectious disease modeling, all of these involved mathematical expertise in an essential way.

The global setting of mathematics now came into being in step with the development of the global economic and political systems that were the heirs of the nineteenth century's great power rivalries and their colonial outreach. The two World Wars led to the redrawing of the world map, as did the collapse of the Soviet Union, with a corresponding blooming and waning of national entities following treaties, the granting of independence to former colonies, and the reconfiguration of central governments into local entities. Yet, despite the proliferation of new nations and ever-shifting borders, the interdependence of the world economy alongside advances in communication technology gradually led to an increasingly interconnected world.

Politically, the period opens with the Paris Peace Conference (1919–20) after the First World War. New states came into being, and along with them new national communities in mathematics, diversifying the international mathematical world and opening new directions for research. The Polish journal *Fundamenta Mathematica*, for example, made set theory a national focus, simultaneously putting Polish mathematics on the international stage. The 1917 Bolshevik Revolution that led to the establishment of the Soviet Union led rapidly to a demand for distinction in this prestigious field. Benito Mussolini's National Fascist Party, which won election in Italy in 1922, sought programmatically to restore Italy to its former glory, thinking not so much of the nineteenth century as of the time of the Medici. Particularly consequential for mathematics was the rise of Adolf Hitler's National Socialist Party in Germany. Appointed chancellor by President Paul von Hindenburg in 1933, Hitler lost no time in carrying out parts of his ideological program, notably with policies directed against Jewish academics, many of whom lost their positions within the course of that year.

German as well as Japanese expansionism led to the Second World War, whose complex outcomes have structured the world stage, including the learned world, ever since. Mathematicians of all sorts participated in these changes. The migration that preceded, accompanied and followed the war, led to a transformation of the mathematical map, as the USA emerged as the single most vibrant research community, benefiting from enormous wealth, newly diversified institutions, and a new determination to be involved in international affairs at the highest level. The new sophistication and increasing complexity of the use of mathematics in research—symbolized by nuclear weapons research, but going far beyond it to embrace all fields of engineering and science—promoted developments in both pure and applied mathematics.

The period after 1945 also saw the disintegration of the British, French, and Japanese empires, with a corresponding rise in demand for the development of national systems in a multitude of new states, including India and China. Many of these states took on the challenge of promoting higher education and research. Britain, France, and the US were important incubators of scientific talent from around the world, as they continue to be. The Soviets competed for talent by creating a special university for African students: the Patrice Lumumba University, named for the assassinated first prime minister of the Republic of Congo. Throughout the former colonial nations, as in the US, Canada, and elsewhere, people of African, South Asian, and Latin origin frequently faced educational discrimination. For example, while US federal law forbade segregation in a series of actions in the 1950s and 1960s, de facto segregation continues to this day.

The rivalry between the West and East (including, after 1949, Mao Zedong's People's Republic of China) manifested itself in the arms race as well as the accompanying competition to conquer space. In 1957, the Soviets launched the first orbiting satellite, Sputnik, which led to deep concerns in the US that the country was falling behind the Russians. This led to a national effort to bolster STEM education—courses in science, technology, engineering, and mathematics. The Cold War was also the locus of an ever-increasing use of the mathematical sciences in military planning and weapons design. Think tanks emerged, like the RAND Corporation, which elaborated scenarios using the new mathematics of game theory (due to John von Neumann among others). An array of smarter weapons soon proliferated, from night-vision scopes to supersonic aircraft invisible to radar. Computers played a key role in these developments, with new generations constantly being rolled out, each smaller and more sophisticated than its predecessor. The lunar landing of Apollo 11 in 1969 highlighted a major propaganda victory for the Americans, not least because it symbolized the capacity of the US to deliver on two fronts: just as it was conquering outer space, the country was also making overall improvements in economic and social conditions, a promise held out by capitalist and socialist systems alike.

Indeed, there was little doubt that the West had clear advantages in delivering human comfort broadly, as well as meaningful forms of freedom, particularly for its elites. While the Soviet Union maintained what prosperity it could—with Russia siphoning off resources from its captive client states—the Americans' strategic exploitation of their superior economic position enabled their clients—especially Germany, Japan, and South Korea—to profit from an increasingly internationalized form of capitalism. Not only the exploitation of know-how and enterprise but also the creation of knowledge itself transformed these societies deeply, putting large-scale pressure on Russia and China. China accommodated more deftly, gradually opening its society to the world economy following the death of Mao in 1976. Mikhail Gorbachev's attempt to reform

FIGURE 0.1. The Pentagon Building in Washington, DC, center of US military power, *c.* 1950. Photo by Charles Phelps Cushing / ClassicStock / Getty Images.

the Soviet Union came too late, however, as the Soviets lost control of their clients, beginning with Poland. The fall of the Berlin Wall in 1989 heralded the collapse of the Soviet regime two years later, the beginning of a nadir period Vladimir Putin and his followers have since then sought to overcome. As we complete this writing, the Russian invasion of Ukraine has already led to massive destruction, in a conflict where both sides make use of some of the most high-tech weaponry available.

Probably the most important changes in the mathematical world over the last three decades, however, have arisen from a more thoroughgoing globalization. The opening of China to the world came with a stupendous economic development that gave rise to a hyper-rich class, who began to send hundreds of thousands of students abroad to study. This went hand in hand with the rapid development of China's educational and research infrastructure, with striking import for the mathematical sciences.

Overall, this period has seen an increasing opening of the doors of mathematics at all levels to the formerly disenfranchised. Important inequities persist, and institutional change has been far from dramatic, but access to higher education around the world has greatly improved, notably for women, even if this often seems to be in a two-steps-forward, one-step-back mode. The imperatives of using the fruits of mathematical research in health, communications, government planning, and many other fields impose a need for basic training up to the secondary level. Moreover, the internet, though still hampered by uneven access and outright censorship, offers tremendous potential for advancing educational opportunities.

Whereas the US and the Soviets acted as the dominant players during the preceding era, with their paternalistic and self-interested sallies into the "Third World," such efforts now continue largely in the background. In the meantime, multinational corporations and state governments (China's in particular) have become a part of that game. At the top of the power scale, government research agencies have heightened rivalries between nations, just as universities have jockeyed for position in international rankings, such as the QS Global Academic Survey or the Shanghai ranking. Less powerful national research communities interact in sophisticated ways with NGOs and other entities to attempt to become more fully participatory in the mathematical world, with varying degrees of success.

The internet, a key tool in this intense expansion and interactivity, came about as a late byproduct of the Cold War, as it arose through a plan originally conceived within the US Department of Defense. Over the last fifteen years, it has pervaded nearly all corners of contemporary life in the wealthier nations, spurred by the invention of the smartphone and the rise of five US companies collectively known today as Big Tech (Google, Apple, Meta, Amazon, and Microsoft). These developments have had profound implications for educational institutions but also for scientific and mathematical research, including the diffusion of the fruits of that research, whether to specialists or the public at large. The original promise of internet democracy inspired gurus such as Richard Stallman to advocate for and enable universal access to tools via open source software and the GNU licensing system. However, the dark side of the web includes criminal activities (pornography, human trafficking, money laundering, and so on), which has led to the development of many national entities to combat them—a leading area of employment for PhD mathematicians. Even more troubling is cyberwarfare, with governments sponsoring illicit efforts to obtain remote control of key infrastructure, like communications and energy transport in enemy states. Such remote control warfare also has a "hot" side, with the running of drones as weapons from safe zones thousands of miles from the place of combat. Mathematics lies at the heart of all this: encryption systems, control of devices, algorithms and sensing all use sophisticated mathematics, and machine learning and artificial intelligence tools are increasingly deployed.

The power of computing has also transformed the very types of problems mathematicians today now attempt to solve. Throughout much of the twentieth century, most applied mathematics began by restricting the phenomena under study to relatively simple models leading to a system of equations with exact solutions. The fact that most real world phenomena are much too complex to fit such a model simply meant that this approach was of limited utility. With the advent of high-speed computing, however, new vistas for applying mathematical methods opened, as it now became possible to find extremely accurate approximate solutions for more complicated scientific models and to do so within a reasonable amount of time. Sparked by military demands, these techniques now combine mathematical and statistical tools for analyzing the mountains of data gathered from a variety of sources in order to draw conclusions for guiding public policy and individual action. In this way, "Big Math" has combined with "Big Data" to become a significant driver of innovation in a number of fields.

This recent trend in research has played a crucial role in efforts to stem the terrifying spread of the COVID-19 global epidemic. If there is any mathematical artifact that everyone around the world has witnessed repeatedly over the last years, it is graphs describing infection rates of various kinds during the COVID pandemic. While these aim to convey important information at a glance, producing such visual aids requires very sophisticated mathematical methods that involve drawing inferences from limited information with amazing speed. The teams of researchers that churn out these graphs do so by combining statistical know-how for sample data with deterministic epidemiological modeling. The latter enables them to predict how the virus will reproduce (in quantitative terms) and what factors will influence that spread (group size, physical distancing). These predictions provide medical authorities with vitally important information when setting guidelines for public policy (stay 2 meters apart, no groups with more than six people, etc.).

This Cultural History of Mathematics approaches the subject in Chapters 1 to 3 by moving upward—from everyday usages to cutting-edge research—and then outward in Chapters 4 through 7—to explore manifestations of mathematics in various other realms of human culture. A key overarching theme is one in which mathematics shifts from a position of being seen as fundamental (philosophically and scientifically, while embedded in the world of learning), to positions in which a range of mathematical sciences has pervaded human activity, such as with infectious disease or climate models. The authors of the seven individual chapters have, each in their own way, offered a portrayal of mathematics as a culturally embedded human endeavor.

NUMBERS FOR EVERYTHING, MATHEMATICS FOR ALL

The COVID example, or indeed any casual look at news media, reveals how numbers—ranging from statistics and indexes to graphical representations of quantitative facts—are today an inescapable part of modern life. The business pages tell us how prices and yields go up and down, often aggregating data to talk of trends past and future. Sports pages do much the same, combining individual and team data to analyze and predict, with implications for gambling that go back centuries, but have now become extremely sophisticated. Governments and industries amass data and use "models" for prediction. Gross domestic product, infection and mortality rates, credit scores, rankings in a dating app—interpreting these and other data-based measures are a part of everyone's lives, all the more so since the mobile smartphone made all of this information accessible and even constructible in your pocket. This availability naturally feeds back into a demand for analysis and for easy-to-use tools that can make all this comprehensible. Very often, these tools have a hidden agenda, perhaps to foster a particular point of view, sell things to a user, or provide useful information to an advertiser.

This contemporary picture emerged from an entirely different situation, as only a century ago arithmetic and a smidgeon of geometry constituted what one meant by general mathematical knowledge (see Chapter 1 of Volume 5). The "universal" education typical of the years up to the First World War was accessible, above all, to those from upper or upwardly mobile classes living in societies fully integrated into the international economy. The transition to the present state of mathematics education has thus involved many new aspects, as Chris Phillips discusses in Chapter 1 of this volume.

The economic evolution of societies has meant an increased role for mathematics in everyday life. Over the course of the twentieth century, an ever-higher percentage of people worked for pay. How much they earned determined how well they could eat, what they could afford to wear, and the quality of the dwelling in which they lived. Even on the most mundane level, basic numeracy required that a person be able to handle their money. Runaway inflation in Germany in the early 1920s, and then the worldwide depression following 1929, underlined the need to cope with figures. Moreover, during this same period, government interventions, for example in economic planning (as in the Soviet Union) or in price supports for agriculture (as in the USA and elsewhere) put a large number of people in contact with numerical and statistical measures for the economy.

The end of the Great War saw an initial effort to integrate the world economy under a set of particular rules monitored and administered by international agencies. The business of these global entities placed a premium on new types of quantitative information. These numbers and indices were, in fact, the very lifeblood of these agencies, the tools they needed in their decision-making processes. These included inflation and unemployment rates, interest rates, and the growth or decline of the Gross Domestic Product (GDP), notions that gradually became standard topics of discussion. This was a new world of numbers that was no longer merely for experts but rather for ordinary educated citizens, since they affected individuals, whether as voters, home buyers or as workers. By the 1930s, thus deep into the Great Depression, policymakers and educators perceived the urgency of getting basic mathematics across to the public, as reflected by such works as Lancelot Hogben's *Mathematics for the Million*, part of a Fabian socialist project praised by Bertrand Russell, Albert Einstein, and H. G. Wells.

As business and industry came to rely increasingly on sophisticated technology, the need for workers with a sound understanding of basic mathematics grew as well. The emergence in the West of mass production, a new feature of industry that gradually spread around the world, frequently required at least a rudimentary level of numeracy on the factory floor. That requirement was even more vital for management. These aspects of modern capitalism were part of a trend in the 1920s associated with "Taylorism," which studied methods for improving the efficiency of industrial production. Managers and workers were now beginning to look at hard numbers in order to assess output and profit. The flipside of this in the centralized economies of the socialist world came in the form of five-year plans, first implemented in the Soviet Union by Joseph Stalin and then two decades later in China under Mao. These policies aimed to raise the GDP in both countries dramatically, thereby transforming the economies of these largely agrarian societies to the benefit of all. The underlying philosophy of these socialist plans was to "lift all boats," in sharp contrast to the capitalist model based on open, often unfriendly competition.

Modern capitalism soon spawned consumerism and the broader notion that "everything has a price." Advertising and public relations soon entered the picture as well. For people working in sales and marketing, charts recording past results alongside future predictions gradually became standard tools. New York City flourished as the original hub of the American advertising industry, which by the 1920s made "Madison Avenue" synonymous with these businesses. Several decades later, sophisticated marketing methods began to take hold within the American political system, a familiar story that has much to do with numbers. In the 1930s, the Gallup organization began publishing opinion polls prior to elections, a practice that required sophisticated sampling techniques. Political polls only became a major obsession, however, during the era of cable television, when viewers grew accustomed to the fact that polling results carried a certain margin of error. Americans also became dimly aware of the role long played by the Nielsen ratings for radio and television programs. This data played a crucial role in the world of commercial media, since it enabled management to set advertising prices based on audience size; it also helped companies reach decisions as to which programs would survive for another season. Ratings and rankings of this sort are no longer new; they have since become firmly embedded in nearly all realms of modern life, above all in social media.

All this simply indicates that numbers long ago pervaded the economic, political, and cultural lives of ordinary people, though often in ways that left most only dimly aware that they were living in a highly mathematical world. The fact that this went largely unnoticed surely had much to do with their exposure to mathematics in schools. In the years following the Second World War, high schools throughout the "developed" world taught everyone basic algebra and geometry, whereas an elite university-bound group learned more advanced material: analytic geometry, logarithms, and trigonometry. Very little of this "stuck" in the minds of most young people once they left school. Nevertheless, those who went on with their scientific and technical education had the chance to reinforce their knowledge of basic mathematics, particularly those parts relevant to course work in their post-secondary studies.

During the period from 1960 to 1990, post-secondary education underwent a veritable explosion. While poorer and less developed countries participated in this only to a lesser degree, it remains nonetheless universally true that much larger portions of the population took up further studies after completing high school. This trend in higher education greatly expanded the base of those exposed to mathematical information or various aspects of statistical methodology. Such people were typically accustomed to unpacking data presented in pie graphs or bar charts; they often also knew how to decipher and interpret the results from polls. Some even understood the language of confidence intervals, which played a major part in quantitative studies in the social and life sciences. Indeed, it was during this period that statistical methods used in the assessment of markets and in the manipulation of public opinion became far more sophisticated.

The 1957 Sputnik launch created a moment when Western educators, particularly in the US, found themselves under pressure to develop a learning culture in which mathematics was not only highly valued but also successfully practiced. While many recognized the need for curricular reform, there was still no unanimity of opinion over what this should look like. Some experts argued that students needed early exposure to the language of modern mathematics, i.e., the concepts and notation of set theory, rather than a thorough grounding in basic arithmetic. This group identified with the "new math" movement, which stressed the need to understand the "true" abstract character of mathematics along with its underlying logical structure. Another camp eventually crystallized in opposition to the proponents of "new math"; this group favored promoting practical applications by emphasizing problem solving as part of the "back to basics" movement.

These debates among educators in Western schools went on for decades, as some remained wedded to older routines, whereas others experimented wildly. In the meantime, however, technology quietly crept into the classroom and soon transformed the terms of the debate. This came about through the development of inexpensive handheld calculators, which literally put a calculating machine in everyone's pocket or handbag. The first pocket calculators from the early 1970s were the property of tech nerds, who quickly threw away their slide rules. By 1980, though, technological improvements (integrated circuits) led to a new generation of devices not much bigger than a credit card. Soon they were in everyone's hands, and people began using them when shopping for groceries, to calculate budgets, or for working on their tax returns. In 1985, Casio introduced its first pocket calculator with graphing capabilities, followed in 1987 and 1988 by Hewlett-Packard and Texas Instruments. The latter could also do basic calculus problems, solving them both numerically and symbolically. These, however, like their far more powerful successor models, remained largely the property of technically savvy users.

By about 1990, then, one could approach much of the standard repertoire of school mathematics by pushing buttons. Not much later, PCs had entered school life in the wealthy world, and young people were learning how to navigate these machines by means of mouse clicks. Many educators in these countries decided to move away from teaching basic skills, convinced that they could somehow make better use of the two to four years previously spent on multiplication and division drills. Yet, teaching problem solving or the fundamentals of modeling without some basic conceptual understanding of mathematics can lead to frustrating experiences. It appears to be the case, in fact, that with or without the acquisition of calculating skills, the conceptual side of the subject often remains murky for nearly everyone. Rules are easier to teach than concepts, but even basic questions taught to schoolchildren during the nineteenth century can leave a modern-day consumer bewildered. Ask someone to explain why a markup of 25 percent, say for a jacket that originally cost $80 and now goes for $100, is not the same percentage when marked down from $100 to $80. Calculating percentages is easy, but understanding what the numbers really mean is far more subtle.

The advent of the internet and the World Wide Web in the mid-1990s produced another seismic shift in the role of numeracy worldwide, a development that intensified after 2005 with the proliferation of smart phones. The latter potentially allowed anyone to have access to vast amounts of scientific information and sophisticated analyses of it made by leading experts. Unfortunately, many users are easily distracted by what is actually only propaganda or disinformation. Moreover, governments can thwart free and open access to the web, and they will do so if intent on controlling what people can read and hear, the Great Firewall of China being the best-known example of this type of containment mechanism.

In the late 1990s, an international movement to promote mathematical literacy began to take hold as part of a broader effort to assess educational achievement in countries around the world. In the year 2000, the Organisation for Economic Co-operation and Development (OECD) introduced the first installment of a worldwide study called Program for International Student Assessment (PISA). In this context, the OECD articulated a clear vision of what numeracy in the twenty-first century should mean: "A mathematically literate student recognizes the role that mathematics plays in the world in order to make well-founded judgments and decisions needed by constructive, engaged and reflective citizens." While many progressive educators have expressed skepticism regarding the PISA studies, a number of efforts both locally and internationally stress the need to put basic mathematics in the hands of everyone. Perhaps the broadest set of initiatives of this kind has attempted to bring girls and women around the world into full contact with mathematics at all levels.

We can also ask what a responsible and mathematically literate person should think about the pressing global problems the world faces today, like the recent pandemic and longstanding changes in the climate of our planet. Do they need to know about how experts deal with data and modeling for the spread of COVID? or the corresponding mathematics needed to predict global warming patterns? While technical experts produce information on a daily basis concerning such matters, teaching how to read and interpret their reports remains a major educational task. Indeed, there is a wide gap between seeing a graphic and being able to understand and judge the commentary and conclusions drawn from it. In some cases, the decisions based on such data and imaging involve life-and-death matters, and in others, they may have profound economic and social consequences. These circumstances suggest a very real urgency to make mathematical literacy widely available in the spirit of the OECD definition. That would mean giving underrepresented countries and peoples the tools needed at least to understand statistical data and the opinions of experts who make use of such studies. Yet, this is happening in the presence of a disarticulation between basic education and research in the mathematical sciences, a phenomenon that older institutions have been slow to grasp.

PROFESSIONALIZATION AND INTERNATIONAL RELATIONS

Just as mathematics became increasingly pervasive throughout modern societies worldwide, so have many nations witnessed a corresponding growth in the number of those directly engaged with higher mathematics. Looking back to the middle of the nineteenth century—the period during which our modern idea of a professional research mathematician first arose—it seems remarkable how tiny that world of "mathematical inventors" really was. Even in Germany, one of the leading countries in the production of mathematical research, around 1900 there were only about 100 full professorships in mathematics at all of the nearly 40 universities and institutes of technology throughout the land. By contrast, a few years ago Jean-Pierre Bourgignon, a French mathematician and national research director, estimated the number of research mathematicians worldwide to be around 80,000. This number becomes much larger, however, if one counts those trained at the PhD level and beyond who work in government and industry. These practitioners are still concentrated in relatively few jurisdictions and it remains overwhelmingly a male-dominated group, despite the growing participation of women at the highest level, and indeed at all levels. Chapter 2, which we co-authored, surveys the vast expansion and partial globalization of the mathematics profession as it evolved over the course of the period 1914 to 2020.

During the latter half of the nineteenth century, European nations largely set the pattern for mathematical research in an era dominated by the rivalry between France and Germany. By the 1920s, however, the United States and the Soviet Union had fast-growing communities, as did Poland, Hungary, Japan, and Sweden. Whereas international contacts between France and Germany came to a standstill in the wake of the Versailles Treaty, new channels opened elsewhere. Regular international conferences in applied mathematics took place, beginning with the one held in Delft in 1924. Paris and Göttingen remained the two leading centers on the Continent, but several others emerged as well, including Rome, Amsterdam, Hamburg, and Frankfurt. The Rockefeller Foundation, founded in 1913 (in the wake of the dissolution of Standard Oil in 1911 as a monopoly) gave a major boost to mathematics through its funding of international exchanges. This permitted networking between institutions at the places mentioned as well as those in Cambridge, Copenhagen, and Princeton.

Up until 1933, Russian mathematicians often spent time at leading centers in the West. Already before the Great War, Nikolai N. Luzin had studied in Paris as well as Göttingen, and he later published regularly in French journals. Much like Chicago's E. H. Moore, who trained the next generation of stars in the United States—Harvard's George D. Birkhoff, Princeton's Oswald Veblen, and Chicago's Leonard Dickson—Luzin laid the groundwork for Moscow's rise to international eminence. One of Luzin's many gifted students was the topologist Pavel Alexandrov, who cultivated friendships with several leading Western European mathematicians. He visited Göttingen nearly every summer starting in 1923 and often offered lecture courses there. Alexandrov was particularly close with the algebraist Emmy Noether, whom he invited to Moscow in the winter of 1928 when she taught a course on modern algebra. In 1933, she was one of the many Jewish mathematicians whom the Nazi government dismissed. Her brother Fritz, a talented applied mathematician who lost his position in Breslau, soon accepted a professorship at the Tomsk State University in Western Siberia.

The Noether family produced four notable mathematicians in three generations, though Emmy stands out among them in view of her deep influence

FIGURE 0.2. Emmy Noether in Göttingen, summer 1932, *left to right*: Otto Schilling, Noether, Olga Taussky, Hans Schwerdtfeger, Ernst Witt, Paul Bernays, unidentified, Erna Bannow, unidentified. Papers of John Todd and Olga Taussky-Todd. By permission of the Caltech Archives.

on modern algebra (Rowe and Koreuber 2020). Considering the diversity of their talents and interests, this single family symbolizes how mathematical research quickly diversified during the early decades of the twentieth century. Although she never held a professorship in Germany, Noether's fame as the "mother of modern algebra" quickly spread around the world, thereby enhancing Göttingen's reputation as a leading international center. By 1933, when she lost the right to teach in Germany, Emmy Noether was at the height of her fame. Thanks to her many friends and influential supporters, she gained a temporary appointment at Bryn Mawr College for women. She and "her girls" (a group of young postdocs) also traveled regularly to Princeton, where Noether offered lectures at the Institute for Advanced Study. As a woman at the pinnacle of research mathematics, in production if not by position, hers was a singular case, though she was one among a small group of women who taught mathematics in institutions of higher education.

Stalin's reign of terror, starting in the mid-1930s, had a chilling effect on Soviet mathematicians, the most visible case being the "Luzin affair," which led to the downfall of the once highly influential Nikolai Luzin (see Chapters 3 and 4). One of the principal charges leveled against Luzin was that he published important work in foreign journals. Recognizing the danger by this time, Soviet mathematicians systematically avoided publishing their work outside the country. None took this matter lightly, as the Great Purge had just begun, yet Soviet mathematics emerged from this period relatively unscathed, unlike the damage done to biological research due to the politically motivated dogma of Lysenko, who had denied the imperative character of genetics.

The exodus of European mathematicians to the United States during the 1930s had profound consequences for the American mathematical community, which had largely cultivated research confined to pure mathematics. Quite a few of its members, to be sure, had engaged in war-related research after the US entered the Great War in 1917, among them Oswald Veblen. As a captain in the army, Veblen oversaw ballistics tests at the Aberdeen Proving Grounds (see Chapter 6), where he later built up a research team during the Second World War. During the intervening period, however, he returned to academic life, continuing his work on differential geometry at Princeton, where interest in Einstein's theory of general relativity spawned much new research on geometrical theories (see Chapter 5). Veblen's case was somewhat typical of the general pattern, although his interests in modern physics placed him in a small elite group, as noted by his friend Birkhoff. When Birkhoff surveyed the landscape of American mathematics over the first fifty years of the American Mathematical Society, he bemoaned the fact that only six researchers in the US qualified as applied mathematicians, and four of these had received their training in Great Britain.

In this respect, the Second World War permanently altered the character of the American mathematical community. Some of its members worked on developing radar systems at Massachusetts Institute of Technology's (MIT) Radiation Laboratory; others, like Stanisław Ulam, joined J. Robert Oppenheimer's research team at Los Alamos working on the production of nuclear weapons. MIT's Vannevar Bush played an instrumental part in launching the Navy's Office of Scientific Research and Development, as well as the Applied Mathematics Panel established under the directorship of Warren Weaver in 1942 (see Chapter 6) . Two former Göttingen figures founded important centers for applied mathematics in the United States: Richard Courant built a mathematical institute at New York University (NYU) practically from the bottom up, whereas Theodore von Kármán created an institute for aerodynamic research at California Institute of Technology.

Following the ascent of the United States, the Soviet Union emerged as the second mathematical "superpower" of the post-1945 era (see Chapters 3 and 4). One of its superstars was Andrei N. Kolmogorov, a pioneering figure in probability theory and dynamical systems whose students made fundamental contributions to these fields of research. Kolmogorov's axiomatization of probabilistic systems in the 1930s was immediately influential in unifying high-end mathematical analysis with the statistical study of data. Mathematical researchers enjoyed a privileged status in Soviet society, nurtured like star athletes and chess masters in a system that cultivated their talents from an early age. Beginning in 1936, the Soviet Union held a yearly Mathematics Olympiad for high school students, a form of competition that served to identify the likely members of the next generation's mathematical elite.

Following the war, the European communities gradually re-established themselves, with continuity or revival in universities and in institutions like the French, German, and Italian mathematical societies. Japan faced the unique challenges associated with military defeat and the necessity of building a scientific system from what had been fragile roots. In this case, international support for the revival of universities and journals manifested itself in foreign visits and research contributions. Alexander Grothendieck, one of the era's leading mathematicians, contributed a landmark paper to the *Tohoku Journal of Mathematics* in 1957. The success of Japanese efforts to promote mathematical research gained further recognition in 1990 when its community hosted the prestigious International Congress of Mathematician (ICM) in Kyoto.

Meanwhile, the new governments in India and China likewise gave prominence to mathematics, as an entry criterion for higher education and as a key component of scientific and technical education. India's first president, Jawaharlal Nehru, regarded the founding of the first Indian Institute of Technology (IIT) in 1950 as symbolic of his country's hopes for a future in which education and research would play a major role. Admission was by competitive examination with mathematics at the core, a tradition that has continued to this day in the IIT's expanded facilities for training highly successful researchers. Here, too, European and American support, both in financial and personal form, played an important part, with the West German government providing substantial funds for the IIT founded in Madras in 1959.

China was at first much influenced by the Soviet model; rather than starting from scratch, the country under Mao's leadership began by building on its older universities in Beijing and Zhejiang. China and India could also point to the achievements of Shiing-Shen Chern and Srinivasan Ramanujan, two celebrated figures in the West, as inspirational examples. Both countries took advantage of their fame by touting these names when they hosted, respectively, the two ICMs held in Beijing (2002) and in Hyderabad (2010). Other states adopted a variety of strategies to develop mathematics, often involving international stars who visited for a limited time: André Weil and Oscar Zariski, for example, spent two years in São Paulo. Decades later, Brazil hosted the 2018 ICM, the first to take place on the continent of South America.

Mention of these "stars" underscores the importance of a kind of star system in mathematics. As in any field, the factors that go into creating star-like status are complicated. Recognition can come from solving, or making major progress on, classically difficult problems, so consecrated by the efforts of earlier masters. Examples include Weil's proof of a theorem related to the Riemann hypothesis—a conjecture from 1859—or Andrew Wiles' success in proving the centuries-old Fermat conjecture. Others have achieved fame through profound or influential innovations or by inventing new methods and approaches, particularly if these prove to be fruitful in an area of hot interest. The Fields Medal, established in 1936 and awarded every four years at the ICMs, represents the most visible form of recognition for outstanding research, being somewhat akin to a "Nobel Prize" for mathematics.

Most recipients, however, did not receive this honor because they solved a famous mathematical problem. Grothendieck, one of the most influential figures of the past century, gained his Fields Medal in 1966 for ideas that redirected and substantially altered ongoing research in several fields. This suggests how influence and fame can sometimes converge, particularly when an important breakthrough serves to guide the course of future research activity. Bill Thurston, a Fields medalist in 1982, profoundly altered the course of topological research by setting forth his vision for classifying three-dimensional manifolds. In the present day, when celebrities often function as influencers through websites on social media platforms, two twenty-first-century stars have emerged as highly active bloggers with large groups of followers: Terry Tao and Tim Gowers, both of whom happen to be Fields medalists.

Since around 1970, research trends and fashions in mathematics have undergone abrupt changes and branching. In the 1960s and 1970s, category theory, point-set topology, and catastrophe theory were all the rage; a decade later they had all but disappeared from the scene as loci of cutting-edge research. Specialization and the urgency to publish new findings quickly have generated an immense wealth of information in recent decades. Yet, even though the overwhelming preponderance of mathematical results arose after *c.* 1950, one should not imagine that this accumulated knowledge simply represents a steadily rising growth curve. A quite different image emerges when one contemplates how quickly mainstream mathematical theories of the last century have receded in importance as libraries are digitized, displacing the print media, in particular the textbook literature, which had long played a central role in the education of young mathematicians.

Seen from this vantage point, the issues of accessibility and retrievability appear in a very different light. For research mathematicians after the Second World War, reference works like *Mathematical Reviews* and *Zentralblatt* became indispensable tools for gaining access to the latest findings in the published literature. More recently, however, the IT revolution has placed massive searchable databases at the fingertips of anyone with a laptop, an internet connection, and access to a subscription. Nevertheless, these are expensive tools, and even in the richer nations, access to them depends on a person's institutional setting. For all practical intents and purposes, professional mathematicians in today's world have only very limited access to the research results of their contemporaries and recent predecessors. Moreover, the collective culture of mathematics has become future-directed to an extremely high degree, paying little if any attention to achievements of even the recent past.

PRODUCING MODERN MATHEMATICS

This volume cannot possibly provide an overview of the many complex changes in mathematical knowledge that took place over the last hundred years or more. Instead, in Chapter 3, Roger Cooke and Leo Corry offer a survey of two contrasting and remarkably influential intellectual traditions that played major roles in creating modern mathematics. These were the Soviet school, with its principal outpost in Moscow, and the group of mainly French mathematicians who belonged to the Bourbaki collective centered in Paris. As suggested above, mathematical invention over the past hundred years was to some extent a seesaw contest between problem solvers and theorists. In an overall sense, the Soviets excelled in the former activity, whereas Bourbaki primarily pursued the latter direction.

During the twentieth century, much pure-mathematical research activity was guided by longstanding major problems, such as the famous 23 that David Hilbert first presented at the Second ICM held in Paris in 1900. Mathematical problems are sometimes quite precise, though in other cases they may have rather broad targets. An example of the former is the Poincaré conjecture (O'Shea 2007), which states that every simply connected closed 3-manifold is essentially the same as a 3-sphere. Whatever exactly this means, it took over a century of work by many people to achieve a positive result, as it turned out that even to approach the problem required developing some highly sophisticated technical machinery. The significance of such problems for the present volume is mainly to underscore the fact that the concerted efforts to solve them often went hand in hand with major theoretical developments. Indeed, theoretical innovation was quite frequently associated with a specific problem, or a set of related problems. In the twentieth century, such innovation in pure mathematics has tended to move away from the concrete world of calculation with numbers. Instead, the focus has shifted to the manipulation of abstract notions constructed to highlight essential features of mathematical objects. For example, a 3-sphere is just a special case of an n-sphere, whereas a 2-sphere is a familiar geometric object, namely the surface of a ball. The abstraction comes about by boosting the dimension and then deciding which features to retain and which to discard.

Throughout the twentieth century, pure mathematicians tended to use the Hilbert problems as a gauge for measuring mathematical progress. Indeed, the two largest national mathematical communities—those in the Soviet Union and the United States of America—published summary accounts of research efforts prompted by these problems some seven decades after Hilbert bequeathed them to his fellow mathematicians. In a similar vein, the Clay Mathematics Institute invoked the memory of Hilbert's feat in 2000 by sponsoring the seven Millennium Prize Problems, each of which carries a million-dollar payoff to anyone who can solve them. To date, only one of these seven problems has succumbed: in 2003, the Russian Grigori Ya. Perel'man solved the Poincaré conjecture. He then created an even greater sensation within the mathematics community by declining both the prize money and the Fields Medal as well (see Chapter 4). Problems in applied mathematics, on the other hand, are often quite concrete: can you model the passage of shock waves through a given base of soil to determine how to make shock absorbers for a 20-storey building? How far apart should people stand in a given circumstance to reduce the chances of catching COVID to 1 in 100? In many cases, such problems have only recently become amenable to mathematical treatment. This depends on whether one can adapt known theoretical structures to the case at hand, but also carry out the computations within a reasonable amount of time.

The boundary between pure and applied mathematics has also blurred considerably in recent years, as collaborative efforts have become increasingly common. Collaboration was very rare during the nineteenth century, in part due to physical separation. Teams of collaborators only slowly surfaced over the century that followed, beginning with the French-based Bourbaki collective, which was perhaps unique in adopting a pseudonym for a group of collaborators (see Chapter 3). Although dispersed, the group had its geographic center in Paris, which often enabled its members to meet rapidly and frequently. At the extreme end of the trend toward collaboration stood the itinerant Hungarian mathematician Paul Erdős, who ranks with Leonard Euler as one of the two most prolific authors of all time with over 1,500 papers, largely co-authored.

The creative works of influential mathematicians are often markedly different. In the period up to about 1980, pure-mathematical fields dominated in the prestige hierarchy, with a corresponding emphasis on the development of abstract theories and tools, both to solve venerable problems and to strike out into new areas of inquiry. As computing came to enhance the power of applied mathematics, alongside parallel developments in probability and statistics, new leaders began to emerge in these areas, which have since grown markedly, with their own prestige hierarchies. Many of these high-prestige researchers were influential both by virtue of having mentored several research students as well as through interventions, charismatic or otherwise, within their community. When pure mathematicians comment on their own work, their remarks often resemble those of creators in the arts, particularly the visual arts or music. Applied mathematicians, on the other hand, often tend to remark on progress in ways that more closely resemble reports by scientists and engineers.

The photographer Mariana Cook produced a book of portraits of some twenty-first century stars, whose comments often provide vivid insights into their mathematical lives (Cook 2009). Drawing on some of these personal reflections, supplemented with brief portraits of others, we here offer a glimpse of the eclectic character of recent mathematical research practices as seen from within. In popular culture, one typically encounters the word genius when speaking about a brilliant mathematician, a label that reminds us of Edison's quip that genius is one percent inspiration and 99 percent perspiration. Several leading mathematicians tend to agree with Edison, downplaying the notion that success in mathematical research involves some special genius. Instead, they describe a special passion, which brings with it the ability to overcome the constant frustrations mathematicians face in coping with inevitable day-to-day failures.

Princeton's Charles Fefferman describes this as a kind of culinary experience: "I don't choose problems; they choose me. A question will grab hold of me and I feel compelled to think about it for years or decades. On a typical day, I get no ideas, but on a good day, I get a wrong idea. Wrong ideas are the ingredients in the pot. Add enough ingredients and the stew cooks. With luck, it tastes good" (Cook 2009: 112). Some leading figures write modestly about their achievements, crediting good luck as a major factor in their successful careers. The Swedish mathematician Lennart Carleson avows to having met a handful of mathematical geniuses during his lifetime, and he confirms that these individuals are indeed very rare. He emphasizes that success in mathematics mainly requires mental tenacity and stubbornness. Carleson spent much of his career working on three problems, devoting five to ten years on each, during which time he thought about little else. Next to pencil and paper, he noted, the most important equipment for a mathematician is a good-sized wastepaper basket. What does success look like? "My experience is that the result slowly dawns on you," says Carleson, "it is only after checking and rechecking that you realize you have eliminated all errors. 'Eureka' is not a very common word for a mathematician" (Cook 2009: 102).

Another popular image of mathematical researchers sees them as lonely brooders, stooped for hours at a time over a problem. Some of today's stars, in fact, have stressed that they prefer to work alone and, in some cases, even outside the realm of mainstream research interests. A dramatic example of this reclusive attitude emerged in 1993 when Andrew Wiles announced he had proved the famous Fermat conjecture. It later emerged that he had spent the previous six years working on this problem in total secrecy.

Still, most mathematicians are anything but loners. Many emphasize the importance of their teachers and/or collaborators and the importance of working with others, whether formally or not. Frances Kirwan represents this larger group by underscoring the importance of collaboration for the success of her

FIGURE 0.3. Even in the twenty-first century, blackboards and chalk remain standard tools in the repertoire of many mathematicians. Photo by Lucidio Studio, Inc. / Getty Images.

career. During the first half of the twentieth century, rather few mathematicians regularly published co-authored papers. Conferences were still relatively infrequent, but in most fields, one could stay abreast of recent developments by reading the journal literature. Mathematicians often refer to the aesthetic pleasures associated with mathematical creativity, though these can take many different forms. Shing-Tung Yau likens mathematicians to painters, some of whom produce abstract works, while others stay closer to the natural world. Working in a closely related field, Sun-Yung Alice Chang compares learning to do mathematics with the discipline required in order to become a professional musician; in both cases, one builds skills by starting with smaller pieces.

One finds references to a particular type of aesthetic experience among self-identified mathematical Platonists, especially number theorists. A striking example is Don Zagier, whose work is classical in that it evinces little affinity with abstract concepts; instead, he tries to explore the deeper implications of concrete, hands-on formulas. Zagier shares the view of the late John Conway that "mathematics is already there and true whether we discover it or not," but he also stresses the subjective side of mathematical research. Mathematicians have different personal styles, and "sometimes just from the statement of a result you can guess which mathematician did the work" (Cook 2009: 162). Zagier thus compares doing mathematics with playing chess, a very common sentiment among pure mathematicians. Chess players follow the same rules, but at a high level of play, the choices made by a grandmaster in a superb game can be innovative and reflect the personal style of that player.

The multitalented prodigy Terence Tao expresses similar views, while sharing an essentially Platonist outlook. Regarding the diversity of mathematical talents, he describes some as sprinters, others as marathoners, and most as slow thinkers. What they have in common is their belief in deeper truths and their willingness to dig for them: "Mathematics can be like archaeology. You might find a corner of something and decide it's of interest. Then you dig somewhere else and find a corner that looks very similar and think there may be some deep connection. You keep digging and finally discover the structure underneath, You have a thrill of discovery when something makes sense" (Cook 2009: 104).

Converts often express interesting views, and Ingrid Daubechies represents one such case: a Platonist turned constructivist. She started her career in pure mathematics, but then began working on wavelets, an important tool in applied mathematics for analyzing digital images, among other things (see Chapter 6). She soon realized that everyone considered these tools to be "constructed" concepts, something made by a mathematician in order to handle a problem. This was very different from the attitude of most pure mathematicians, and she, too, had shared the same sense they did of "discovery": what one found in mathematics was somehow already there. Moreover, she felt no differently with regard to her work on wavelets, which made her ponder where the boundary between these two realms might lie. Eventually she concluded that there really is no such boundary and that all our mathematics is constructed.

Other mathematicians express themselves on the esthetics of mathematics. The best known of these is likely the English mathematician Godfrey H. Hardy, who in his 1940 essay *A Mathematician's Apology* stated that there is no room in the world for ugly mathematics. Hardy proceeded to insist that the theory of numbers—one of his own main areas—epitomized the beauty of mathematics precisely because of its lack of applicability. Such comments have made generations of applied mathematicians bristle, decrying such criteria as both artificial and simply wrong. Such debates over esthetics are not resolvable, but the point here is that there are many different points of view regarding what is good or beautiful mathematics. These play out often in the committee discussions leading to prizes.

Who has the opportunity to produce modern mathematics? The biographical commonalities and disparities of the illustrious people we have just mentioned are not easy to link with their mathematical careers. One point they had in common, though, was access to elite education, whether through a national structure or through lucky circumstance. This factor alone certainly helps to account for the longstanding underrepresentation of women in mathematics. It looms even larger, however, when considering the limited participation of people from disadvantaged minorities. Indeed, those who grow up in countries whose educational systems provide few opportunities for advanced study—whether in Africa, South America, or large parts of Asia—clearly stand at a tremendous disadvantage. That alone is reason to look at a few exceptional cases of people from such backgrounds, who nevertheless managed to make their way into the higher echelons of the mathematical community by virtue of the excellence of their creative work.

David Blackwell was the first African American to become a tenured professor at the University of California, Berkeley, one of the leading mathematical centers in the world. Blackwell was born in 1919 in a small town in Illinois, where he attended a racially integrated school. A precocious child, he excelled academically, completing high school when he was only sixteen. He then went on to study at the University of Illinois in Urbana-Champagne, hoping to become an elementary school teacher. Along the way, a young faculty member, Joseph Leo Doob, discovered this talented Black student and agreed to supervise his doctoral work. By 1941, when Blackwell had just reached the age of 22, he already had his PhD.

Joseph Doob grew up in a Jewish family in New York. Doob's interests were far from the mathematical mainstream of his day, but his work eventually exerted a strong impact by bringing rigorous mathematics to inferential statistics. In his 1934 paper, "Probability and Statistics," he used the new probability theory of Kolmogorov to produce airtight proofs for major statistical results. Some statisticians resisted this line of thought, but eventually it became the dominant approach to the field, making Doob an influential and prestigious figure. The *Mathematics Genealogy* website shows that Doob has had over 1800 "descendants," meaning people who obtained PhDs from him, one of his students, or one of their students, etc. Almost 500 of these happen also to have been descendants of David Blackwell.

After taking his degree, Blackwell did postdoctoral work at the Institute for Advanced Study in 1941, but he could not attend lectures at Princeton University because of his race. In 1944, he gained a professorship at Howard University, a venerable, traditionally black school in Washington, DC. After a stint doing statistical research at the RAND Corporation in the late 1940s, Blackwell became a visiting professor at UC Berkeley in 1954. One year later, finally overcoming racist obstacles, he gained a permanent appointment in Berkeley's new Statistics Department. Blackwell later became the first African-American ever elected to the US National Academy of Sciences.

The life of Maryam Mirzakhani, the first woman to win the Fields Medal, presents another kind of exceptionalism. Born in Iran in 1977, she was still very young when the Islamic revolution took place. Thus, Mirzakhani grew up in a traditional Shiite society, where women adhered to strict traditional customs, such as wearing the hijab. Iran, though, has a strong system of mathematics education, including training of the most capable for competitive problem solving. Mirzakhani excelled at these competitions, and she stunned the world in 1995 by getting a perfect score and a gold medal at the International Mathematics Olympiad. By this time, she was becoming internationally renowned in mathematics, having begun to publish her work while still an undergraduate at Sharif University. Mirzakhani left Iran to take a doctorate at Harvard, which she completed in 2004. She spent the next four years at the Clay Mathematics Institute and in Princeton, before moving to Stanford in 2009. Her Fields Medal came in 2014, though she died of cancer three years later. Only those few who really understand Mirzakhani's work can appreciate her mathematical influence—a typical situation for the creative mathematician today—but her life story has inspired many others as well. Her name and fame as a woman from the "Third World" who went on to attain the highest honor—since joined in 2022 by the Ukrainian Maryna Viazovska—has been enlisted often in calls for the removal of educational and career barriers that have stood in the way of women for centuries.

Another notable product of the Iranian system was Caucher Birkar, who received a Fields Medal at the ICM held in Rio in 2018. An ethnic Kurd, he grew up on a subsistence farm in the province of Kurdistan and later studied mathematics at the University of Tehran. His first breakthrough came in 2000 when he won third prize in the International Mathematics Competition for University Students. Soon thereafter, he sought political asylum in the UK and began studying for a PhD at the University of Nottingham. Soon after emigrating, he adopted the name Caucher Birkar, which means "migrant mathematician" in Kurdish. Today he is a Fellow of the Royal Society and holds professorships at Cambridge University as well as at Tsinghua University in Beijing.

MATHEMATICAL WORLDVIEWS FROM FOUNDATIONS TO TOOLS

Hilbert's distinctive approach to mathematics was part of the large current of research using axiomatization as a central tool that emerged in the 1890s (see Chapter 3 in Volume 5). Among other things, the creation of non-Euclidean geometry had elevated awareness of the importance of basic mathematical assumptions—axioms—in the construction of mathematical theories. Giuseppe Peano and other leading Italian mathematicians stood at the forefront of this movement, which received significant attention in Hilbert's 1900 Paris lecture "Mathematical Problems." Particularly crucial was his second problem, which called for proving the consistency of the axioms for arithmetic. This means one must show that it is impossible to derive two results from these axioms that contradict one another. Proving that something is impossible is (obviously) not easy. The upshot of this second Paris problem was the creation of a formalist research program, which Hilbert launched in earnest soon after the Great War ended (see Chapter 4).

One major fruit of this formalist approach was proof theory, created by Hilbert and his assistant Paul Bernays in hopes of using mathematical tools to establish the consistency of arithmetic. This effort dominated their attention throughout the 1920s, but it ultimately failed, at least in the original form in which Hilbert had conceived his theory. In 1930, the Viennese logician Kurt Gödel turned proof theory on its head by showing that the consistency of arithmetic was a formally undecidable proposition. This meant one could neither prove nor disprove it within the established system of axioms. Thereafter, Hilbert's proof-theoretic methods triumphed, even though his optimistic mathematical worldview had been largely shattered. His larger goal, in fact, aimed to dispel all doubts as to the soundness of everything in mathematics, including Cantor's highly controversial theory of infinite sets.

Hilbert seemed to have abandoned the second Paris problem after 1905, but in 1917, he picked up again where he had left off. This set the stage for a dramatic clash of mathematical titans, a storm that had been brewing for some time. Already in his doctoral dissertation from 1907, the Dutch topologist Luitzen E. J. Brouwer had leveled sharp criticism at Hilbert's axiomatic methods, especially his claim that the mathematical existence of a concept was equivalent to saying that its properties are not contradictory. This stance was unlikely to please everyone; would it be enough, for example, to prove the existence of God?

Brouwer's views, labeled "intuitionism," were not widely known prior to 1920, the year in which he began prosecuting his campaign to reform mathematics. The term intuitionism was inherently ambiguous, and those who identified with it held diverse views about the basis of mathematics. Hermann Weyl, Hilbert's most gifted disciple, leaped headlong into this simmering conflict when he published his manifesto "On the New Foundations Crisis in Mathematics." Some historians—in particular Herbert Mehrtens, Paul Forman, and Reinhard Siegmund-Schultze—have argued that the divide separating Hilbert and Brouwer reflected other factors, especially the backdrop of cultural pessimism in the Weimar era, a mood reflected in the intellectual discourse of physicists as well as mathematicians (on debates over quantum physics, see Chapter 5). Indeed, Weyl penned his manifesto during the period of the abortive Kapp *Putsch* that had briefly threatened to end the Weimar Republic, a circumstance that helps to explain why he supported his position by means of political and economic metaphors.

In Chapter 4, José Ferreirós describes how this dispute reflected a power struggle between Brouwer and Hilbert, one that hinged on their conflicting views regarding existence proofs in mathematics. Hilbert claimed that, in mathematics, every well-formulated problem has an answer, or expressed in logical terms, a proposition A is either true or its negation—A is true. This is the classical law of the excluded middle (*tertium non datur*—a third [possibility] is not given). In fact, Hilbert even claimed that every meaningful question in mathematics should be *decidable*—that is, mathematicians can determine whether it is true or false in a given system of consistent axioms. Brouwer rejected the universal validity of the law of the excluded middle, claiming that this amounted to Hilbert's dubious claim that every mathematical problem has an answer. For Brouwer, mathematical knowledge was time dependent, not transcendent, which meant that it was illegitimate to presume a proposition was knowable.

This is perhaps as good a place as any to mention that, while mathematics takes place within societies, mathematicians themselves not infrequently regard this as entirely irrelevant. Indeed, the apparently transcendent character of mathematical knowledge ($2 + 2 = 4$ no matter what) often leads mathematicians to attribute a kind of independence of their own work from societal events. Some would go so far as to claim that mathematics has no political or social responsibility whatsoever, like many other academic fields, though typical justifications for ivory tower attitudes in mathematics are no doubt unique. In the present case, however, the importance of social and political background factors was evident. Brouwer became increasingly engaged in stirring up German animosity against the French—an attitude that gained him many allies following the Great War. His campaign effectively split the German mathematical community along political lines, leading Hilbert to oust Brouwer from the editorial board of *Mathematische Annalen* in 1928 (for details, see Chapter 4).

These events aside, the philosophical world, and even informed public opinion, took far more interest in the revolutionary aspects of physical thought that came to the fore in the 1920s. During the interwar years, Einstein and relativity became synonymous with a new understanding of the profound mysteries of the universe (Rowe and Schulmann 2007: 97–137). Although hailed in the popular press as the "new Copernicus," Einstein rejected the notion that his theory of relativity represented a new worldview comparable to the heliocentric system that Galileo defended until brought before a tribunal of the Catholic Inquisition in Rome. He was nevertheless eager to impart the main ideas behind relativity to a larger public, and he did so very effectively in *Relativity: The Special and General Theory*. This little booklet appeared in 1917 and soon after the war became a scientific bestseller, sparking numerous debates between various friends and foes of relativity, few of whom possessed the competence to understand its deeper issues.

Hilbert and Einstein lived during an era when mathematics was central for both logical and physical foundationalism. The idea of a coherent worldview, though, is at odds with the eclecticism that now rules in the sciences and in philosophy. If mathematics around the time of the Great War appeared to have a foundational character, linking logic to reality and philosophy to physics, the role of mathematics in worldviews today, and the role of worldviews for mathematicians, are highly heterogeneous, in ways that go far beyond the disagreements between formalists and intuitionists. The banning of metaphysics, associated with the formalist Vienna circle, is now highly unfashionable in academic philosophy. Even more, no one in educated society thinks of "everything" as issuing from mathematical premises.

While the Bourbaki project lives on, it seems clear that there will never be a unified approach or uniform set of definitions for all of mathematics so long as mathematical knowledge remains in human hands. The different parts of pure mathematics may, and often do, interact, with tools from one field (like topology or probability) adapted to another (like combinatorics). Applications, on the other hand, tend to develop from external sets of concerns. These may include both mathematical issues (like the order of complexity of a computation, or the convergence of a method) as well as more remote concerns linked to the area of application. In particular, one judges the success of a model by scientific, practical, or even commercial criteria. In this context, metaphysical issues are wholly absent, whereas deeper formal issues are likely to be either invisible or largely irrelevant, in part because the methods will remain useful no matter how they are justified at a foundational level.

Instead, ethical issues have since taken on growing importance. One area in which this has occurred is when considering the underlying assumptions of algorithms that operate as contemporary tools in artificial intelligence. These algorithms assist in carrying out parts of decision-making processes, as when physicians use computers to make a medical diagnosis, or (to take a more familiar example) when you choose a film on Netflix because "others who like this watched . . ." Since the creators of these programs are very often young men, who come from a restricted range of educational and cultural backgrounds, it is hardly surprising that studies have shown the influence of those backgrounds in the algorithms they design.

The enormous diversification of approaches to mathematics has resulted, in fact, in the application of a wide variety of tools from across the mathematical sciences. Climate change provides a prominent challenge here, starting with the need to predict the future impact of CO_2 emissions. Scientifically grounded information about the potential consequences of global warming should play an important role in public conversations about this problem. This includes assessing the potential impact of political and economic measures aimed at curbing the effects of rapid climate change. While equations describing the short-term evolution of the atmosphere—as potential tools for weather forecasting—have long been known, solving these still poses major difficulties. Even small-scale weather prediction requires crunching enormous amounts of data, and climatologists still do not completely understand such fundamental phenomena as cloud formation. How much more complex, then, is modeling the climate? These studies of the atmosphere are part of the mathematical field known as computational fluid dynamics, which involves combining a host of deterministic and statistical models with differing degrees of confidence in the outcomes. This field of research clearly poses major technical challenges, but that is not all; the results obtained also require translation into a language others can understand. This requires parsing and assembling the findings in such a way that they become both accessible and meaningful to policy makers, journalists, and the wider public. As with COVID modeling, studies of climate change rely on a heterogeneous set of methods, sharp understanding of their reliability, and an ability to communicate all this convincingly. These challenges have created, in effect, an entirely new role for mathematical scientists in the world of the early twenty-first century, and their work underscores the place of mathematics in today's worldviews.

GENERAL RELATIVITY AND NUCLEAR PHYSICS

While philosophers and the philosophically inclined literate public have shown interest in the ways mathematics informs our understanding of the physical world, the actual interplay between mathematical ideas and modern physics during the last hundred years has been truly profound. On a technical level, new mathematical methods have vastly extended the realm of problems physicists can both address and solve. At the same time, on the conceptual level, mathematical theories of recent vintage (such as group representation theory, a method for expressing naturally occurring symmetry in a mathematical form amenable to computation, as noted in Chapter 5) have developed hand-in-hand with important new physical theories (such as elementary particle physics). This has created a dynamic in which all theoretical physics uses mathematical concepts, some rather abstruse, and where applications routinely involve mathematical modeling of physical processes coupled with empirical testing of the model. For continuous processes, the usual tool consists of a system of partial differential equations. A process modeled by equations then describes the evolution of a physical system, but with simplifying assumptions that limit the range of application of the model. Solving the equations provides a picture of what the system should look like in the future. This sketchy description belies, however, the enormous mathematical difficulties that typically arise.

Einstein's gravitational field equations, for example, are notoriously difficult to solve, so he and others devised approximation methods for handling physical problems of special interest. He then used these for calculations connected with testing general relativity against astronomical observation. Already in December 1915, the German astronomer Karl Schwarzschild made an important mathematical breakthrough while stationed on the Russian front. Schwarzschild

FIGURE 0.4. An artist's impression of a black-hole accretion disk (NASA). Wikimedia Commons.

gave the first exact solution of Einstein's equations by deriving the gravitational field outside an imaginary "sun"—a non-rotating homogeneous sphere of given mass. This so-called Schwarzschild solution allowed for precise calculations of the tiny deviations from Newton's theory that were crucial for general relativity. Schwarzschild also noted there was an interior solution as well, although he and other experts attached no physical significance to it. This solution implied a "Schwarzschild radius," which astronomers long believed acted as a barrier beyond which light from outside could not pass. However, via sophisticated mathematics and a modification of assumptions, astrophysicists eventually realized that heavy stars could collapse inside this radius, leading to a black hole: a celestial object from which light could never escape; it only allowed for incoming, but not outgoing traffic (see Chapter 5).

No one will ever see a black hole, though artists have often depicted what such phenomena might look like. By the 1960s, Stephen Hawking and Roger Penrose already set forth strong mathematical arguments for presuming their existence. Penrose introduced the concept of trapped surfaces on the way to proving that spacetime singularities—points where the field is no longer defined,

FIGURE 0.5. Sir Roger Penrose, May 1992. Photo by Louis MONIER / Gamma-Rapho / Getty Images.

like dividing by zero—must exist under quite general physical conditions. Since then general relativists and astrophysicists have fully realized these unanticipated consequences of general relativity, as recognized by the Nobel Committee that awarded the 2020 Nobel Prize in Physics to Roger Penrose and two astrophysicists, Reinhard Genzel and Andrea Ghez, for showing that a supermassive black hole governs the stars' orbits at the center of our galaxy. Only a few years before, physicists finally confirmed a much earlier prediction of general relativity, namely the existence of gravitational waves, a conclusion Einstein had drawn back in 1918.

Along with the theory of relativity, Einstein's name has often been associated with the first nuclear weapons. Already in 1905, Einstein had derived the theoretical equivalence of mass and energy from the special theory of relativity. Forty years later, the latent power expressed in the famous formula $E = mc^2$ revealed itself to the world in the destruction of Hiroshima and Nagasaki. Mainstream media quickly seized on Einstein's mysterious formula in dubbing him "grandfather" of the bomb (Oppenheimer being the "father"). Although Einstein was then the world's most famous physicist and an outspoken anti-Nazi, he had little idea of the US nuclear program. Moreover, it was not until the late 1930s that physicists began to imagine the relevance of $E = mc^2$ for the nuclear realm. By that time, physicists were only beginning to unlock the mysteries hidden within the atom.

A first breakthrough occurred when German physicists Otto Hahn and Fritz Strassmann identified barium as a byproduct of such a bombardment of uranium, a result interpreted as nuclear fission by the Austrian physicist Lise Meitner. Word of this discovery quickly reached Copenhagen, just as Niels Bohr was about to depart for a lengthy stay at Princeton's Institute for Advanced Study. Soon after his arrival, Bohr, working in collaboration with John A. Wheeler, realized that only the uranium-235 isotope could be split by slow neutrons. This represented a major setback since U-235 accounts for less than one percent of natural uranium; moreover, the technological problem of separating isotopes appeared arduous, perhaps not even feasible. The eventual Manhattan Project, combining work of mathematicians, physicists and engineers, overcame a variety of theoretical and technical problems, providing a dramatic demonstration of what such teamwork in science could achieve, and creating a demand for more.

The physical core of these developments, however, depended on quantum mechanics. As explained by Marta Jordi Taltavull and Tilman Sauer in Chapter 5, during the 1920s quantum mechanics emerged as the new physical orthodoxy for describing phenomena at the subatomic scale. These incorporate chance in an essential way, demanding that the physicists' mindset move away from the determinism of Newtonian mechanics. In the classical world, knowing the mass, position and momentum of a particle at one point in time should allow us to know where it will be at any point in the future. Yet, in the quantum world, one knows only probabilities, with the outcome decided only when measurement occurs. The quantum viewpoint also required the making of a new mathematical formalism. Hilbert was involved here, too, though John von Neumann carried out the elaboration of this theory. In fact, two equivalent formalisms—matrix and wave mechanics—emerged to account for the same theoretical effects.

The inherent difficulties with a probabilistic interpretation of outcomes, which in the older intuition were deterministic, was not the only problem brought forth by the new theory. The uncertainty principle stipulated that the position *and* momentum of a particle can never be determined at the same time, because finding one of these means that the other can only be specified as lying within a certain interval. While Bohr's Copenhagen Interpretation attempted to provide a coherent philosophical alternative to Newtonian determinism, as time went by most physicists and users of physical theory became content with a method that allowed the stable prediction of outcomes. Some still strove to work out a conceptual basis for this, but for many quantum theory uncoupled the link with a deep reality, thereby freeing physicists from the notion that their models provided a perfect mirror of the phenomena they studied.

Quantum orthodoxy has continued to preside over developments in subatomic physics, despite certain highly unintuitive conceptions (like mass renormalization, which treats infinity as finite in certain cases). The continued success of this set of programs now pervades physics from cosmology to foundational physics, as well as most applied parts of materials science and optics. One signal success that received Nobel recognition was the work of Steven Weinberg and Sheldon Glashow, proposing that the four fundamental forces now observable (gravity, electromagnetism, and the strong and weak nuclear forces) had been only three in the very early universe. This entailed the existence of three "bosons," heavy subatomic particles, which have in the meantime all been detected, the last—the Higgs boson—confirmed in 2013. These are objects whose existence followed by virtue of a mathematical theory, and this "standard model" proved indeed to be a faithful model of a portion of the physical world. Weinberg himself felt gratified by the success of these tests, but he remained interested in further tests, notably those associated with quantum computing.

MATHEMATICS AND COMPUTING: A SYMBIOTIC RELATIONSHIP

The pure mathematics of the twentieth century often seems far from what most people think of as the core of mathematics, namely arithmetic computation. Over the course of the period, in fact, the importance of computational mathematics grew markedly, leading to a variety of efforts to speed up calculations and reduce error. While the electronic digital computer arrived too late to support the needs of the Manhattan project, soon afterward it signaled what would become a massive shift for mathematicians and those who used mathematics in applications. These developments also fed back into mathematics, following a direction based on logic and Hilbertian formalism that was already well underway before the war began. Alan Turing, a key figure here as well as in wartime cryptanalysis, stood at the cutting edge of these efforts, which sought to determine the essential features of computation by articulating what

FIGURE 0.6. Howard Aiken in 1944 with the IBM Mark I, presented to Harvard University for research by the US Navy during WWII. Photo by Bettmann / Getty Images.

it means for something to be computable. For the purposes of this investigation, he came up with the notion today known as a "Turing machine." Such an idealized device gained considerable interest once real machines arrived that could carry out large numbers of calculations with speeds far exceeding human capacity. In Chapter 6, Liesbeth De Mol discusses many aspects of these developments and their subsequent importance for mathematics.

California was one important nexus for events in the mathematics of computation and discrete mathematics, illustrating the vitality of these research directions. Olga Taussky and John Todd were instrumental in creating the Institute for Numerical Analysis, based at the University of California, Los Angeles (UCLA) and funded by the US National Bureau of Standards from 1947. Another important early outpost for research in computational and discrete mathematics was the University of California in Berkeley, where Derrick H. Lehmer worked with his student Ronald Graham. A highly original mathematician, Lehmer had studied at Brown under the Russian emigré Jacob D. Tamarkin and spent time at both the IAS in Princeton and in England making a wide range of mathematical contacts. After taking his doctorate, Lehmer moved to California (Cal Tech, then Berkeley), returning east in 1945 to work for the Computations Committee at the Aberdeen Proving Ground (see Chapter 6). There, he was part of the group charged with developing ENIAC, the computer created at the University of Pennsylvania during the closing years of the war. In 1943, Lehmer founded, along with Raymond C. Archibald, the journal *Mathematical Tables and other aids to Computation*, a publication of the National Research Council in response to wartime requirements. As a pioneer in non-electronic number-theoretic calculations, Lehmer rapidly adapted to the electronic world. During the McCarthy period, however, Berkeley fired him for refusing to take a Loyalty Oath, though the administration later reinstated him in 1952.

The Cold War era also played a part in shaping Graham's early work. Born in California in 1935, Ron Graham took his doctorate at UC Berkeley under the direction of Lehmer in 1962. Afterward, he immediately gained an appointment as Director of Information Sciences at Bell Laboratories (later Bell Core), where he served as Chief Scientist from 1995 to 1999. Bell Labs, long part of what began as a telephone company, enjoyed an astonishing record of innovative technical achievements. An important breakthrough came with the invention of the transistor, a key component of the digital computer that replaced vacuum tubes. Then followed the silicon solar cell; the first patent for a laser; the first communications satellites; the theory and development of digital communications; and the first cellular telephone systems; the first fiber optic cable systems; and also Unix and C, the operating system and programming language, respectively. The importance of new information sciences to these endeavors put Graham in a position both to create and monitor promising new research as well as to oversee and cultivate applications. The fact that Bell included the mathematics of computation so deeply in its plans illustrates a new awareness of the importance of discrete mathematics, an importance that was to grow as these areas developed.

This terrain involves what are often called "counting problems." Most mainstream mathematicians of the mid-century regarded such problems—often easily posed and yet hard to solve—as curious puzzles rather than serious topics for researchers. By virtue of their simplicity and concreteness, they saw counting problems as akin to recreational mathematics. Such work did attract the attention of excellent mathematicians, however, one of them a legendary figure, Paul Erdős. After fleeing Hungary for the United States in the wake of Hitler's rise to power, Erdös spent his life going from one conference or research center to another, living out of a suitcase, while posing and solving literally thousands of problems in discrete mathematics. His joint publications—over 500—were such that the Erdős number of a mathematician counts the number of steps between a person's collaborators and Erdős himself. Ron Graham, who met

Paul Erdős in 1963, wrote over 30 papers with him. The itinerant Erdős eventually moved in with Graham and Fan Chung, Graham's collaborator, co-worker at Bell Labs, and wife; he lived with them (when not traveling) up until his death in 1996. Two years later, Graham and Chung published their book *Erdős on Graphs*.

Chung, born in 1949, did her undergraduate work at the National University in Taiwan, which was under Japanese rule up until 1945. Despite the then conventional notion that higher mathematics was an exclusively male domain, four female students there finished at the top of their class in 1970, and all four went on to have stellar careers: Sun-Yung Alice Chang, Wen-Ching Winnie Li, Fan Chung, and Jang-Mei Wu. After graduation, Chung began graduate studies at the University of Pennsylvania, where she took her PhD in 1972, afterward going directly to Bell Labs. There she found many collaborators, among them Graham, producing a diverse portfolio of research centered on graph theory and related fields. Chung and Graham both divorced earlier partners before marrying in 1983. Their lives thus reflect changing social norms for academic couples—during the early decades of the century such a relationship would have caused a major scandal.

Computer science is not, of course, all about mathematics, and the spectacular growth of the field led rather quickly to its spinning off institutionally into a separate discipline. Moreover, fast-paced technical developments and numerous commercial links produced a distinctive academic culture quite different from the slower world of traditional mathematics. Moreover, the consequences of computing for mathematics went far beyond stimulating established subfields and spawning new ones. By the year 2000, many once marginal areas of discrete mathematics—graphs and networks, combinatorics, the analysis of algorithms, operations research—had come into their own. They not only had their own journals and awards as well as their own consecrated classical problems, they were also developing their own centers and networks for research and teaching. Much as analysis interfaced with physics in the previous century, these fields have decisively shaped theoretical and practical approaches to a wide range of new areas that have emerged during the twenty-first century. They have played a major role not only in computing, but also in economic and business analysis as well as in solving engineering problems of various kinds, not to mention traditional scientific applications.

Perhaps because these newer areas of mathematics were fresh and little explored, they enabled young people to reach the frontiers of creative research activity without having to invest several years studying the scholarly literature. This may be part of the reason these fields have attracted students who were formerly outsiders, including women. The careers of Graham and Chung also illustrate the increased opportunities for participation in mathematical research outside academia. Today thousands of employers use the skills of mathematicians in fields from security analysis to telecommunications and systems engineering.

MATHEMATICS IN LITERATURE AND MODERN ART

Throughout the 1920s, Einstein's fame continued to grow and soon exceeded by far that of any other scientific figure from that era. Journalists found him highly photogenic and by no means averse to cultivating adulation; he was also a worldly genius, who spoke out willingly and forthrightly about human affairs. His appeal thus extended to a number of different groups, whereas others, especially anti-Semites, generally loathed him and his cult-like following. For many Americans, Einstein stood as a foremost symbol of scientific and social progress. Yet, even for those who saw the impending modern age as a threat, he held a special place in the pantheon of scientific heroes. For in some ways, the broader interest in relativity harkened back to the earlier public discourse concerning the geometry of space, beginning with the publication of works by Riemann and Helmholtz (see the Introduction to Volume 5).

In one of his poems from *Viva* (1931), e e cummings took up Einstein, drew on Robert Frost and Friedrich Nietzsche, and summed up the lament of humanist intellectuals who were instinctively repelled by modern science and technology. Cummings steps into the role of a befuddled lecturer who gets off to a bad start by declaring: "Space being (don't forget to remember) Curved"—and then keeps losing his train of thought. Quite apart from its innovative style, the poem alludes to recent and past scientific theorizing by adopting a mocking tone. With stumbling allusions to famous achievements, cummings brings us quickly to his main message: the crowning of the new "Lord of Creation, MAN." Yet, Einstein had preserved Newton's law—perhaps a symbol for nature's timeless abode—and the "Lord of Creation"? He displays his power over nature by "compassionately" shooting elephants, blowing them to smithereens, while the "electromagnetic continuum" vanishes, and MAN returns to the mechanistic laws of a billiard ball universe. Cummings clearly had his doubts about this brave new world, not unlike William Blake's sense of repulsion in contemplating the implications of Newton's soulless vision.

In Chapter 7, Nina Engelhardt describes how mathematical motifs took on a variety of meanings in major literary works of the last century, many of which reflect tensions between rationality and human freedom. A number of these portray the latter as threatened by the former, but many others take a more nuanced view. In George Orwell's famous dystopian novel *Nineteen Eighty-Four* (1948), the protagonist Winston Smith attempts to hold on to a simple mathematical truth in a society that manipulates facts, memories, and perceptions to suit the purposes of its leaders. In this repressive system, so antithetical to personal liberty, Winston hangs on to a simple fact that symbolizes a truth the totalitarian Party cannot deny: "Freedom is the freedom to say that two plus two make four."

Many modern artists also took inspiration from various manifestations of mathematics, particularly those relating to geometric forms. As discussed in Volume 5, the heyday of geometric model makers ended long before the outbreak of the Great War. By the 1930s, the leading trends in mathematical research were moving in the direction of abstract theories and away from earlier interest in concrete objects. What mathematicians thought of as concrete, on the other hand, proved highly inspirational for abstract artists. Popular discourse concerning four-dimensional space and non-Euclidean geometry rippled through the Dadaist and Surrealist movements as reflected in many works from this period (Henderson 1983/2014). In the mid-1930s, Max Ernst took his friend Man Ray to the Institut Henri Poincaré (IHP) in Paris, where the models on display had been gathering dust for some time. Ernst began placing drawings of these in his collages, whereas Man Ray produced a series of photographs of the models designed to enhance their aesthetic dimension. Using various lighting effects and an opaque indeterminate background, he effectively turned these models into what Surrealists called "found objects."

FIGURE 0.7. Man Ray, Mathematical Object, 1936, photograph of model made in 1879 by Karl Rohn. © Man Ray 2015 Trust / DACS, London 2022.

These photographs went on display in Paris in 1936, and later in London and New York. Man Ray left France in a hurry after the outbreak of the Second World War and would not see his photos again for several years. Having relocated to Hollywood, he managed to retrieve many of his works in 1947. Between 1947 and 1954, he then produced a series of paintings based on the photographs, which he called "Shakespearean Equations."

One of the objects that Man Ray photographed was a plaster model of a special quartic surface originally constructed by the German geometer Karl Rohn (see Figure 0.7). Connoisseurs of modern art might recognize Rohn's model—either from the photo or from Man Ray's painting *King Lear*—whereas very few mathematicians would be able to answer any of the obvious questions, such as who made this model and what does it represent, or when was it made and why? This model represents one of the three types of Kummer surfaces Rohn designed in the late 1870s; afterward copies were sold to universities throughout Europe and the United States (see Chapter 7 in Volume 5).

This example of a "found object" raises a more general question though, namely, when does mathematics become art? Ernst E. Kummer made models of special surfaces for private use, whereas the Rohn models were sold to institutions and mathematicians around the world. Man Ray was surely not the first to find aesthetic value in certain older geometrical models, even if his interest in them was entirely non-mathematical. Surrealist artists presumably had no meaningful interactions with mathematicians, not that they would have had anything to learn from them. After all, those who excelled in model making, like Rohn, had long since passed from the scene. Gaston Darboux had apparently once used the models at the IHP in courses he taught at the Sorbonne, but who would have remembered that in the 1930s? What we can say with assurance is that the younger generation of "pure" mathematicians in Paris took no interest in these special objects, which held out nothing of relevance for their far more general theories. Geometers had become topologists and, for them, there was no reason to fixate on what goes on in a space of dimension 3, especially if the very same thing occurs in dimension n. By contrast, Man Ray's eyes were fixed on the concrete; he took inspiration from artifacts he and others associated with a mysterious mathematical realm. Shakespearean Equations thus represents an artist's attempt to "humanize" mathematics.

In the case of the Dutch graphic artist, Maurits C. Escher, the interplay with mathematical phenomena was far more direct and obvious, though many of his most famous drawings conveyed optical illusions. Escher had no special training in mathematics, but his works evince an intuitive understanding of geometrical structures and symmetries. In 1922, he went to Italy and Spain on a study tour that provided much inspiration for his early work. In Granada, he examined the intricate decorative designs of the fourteenth-century Alhambra with their interlocking repetitive patterns. Later, Escher's brother informed him that the mathematician Georg Polyá had analyzed such patterns systematically, and thereafter tessellations of the plane or the interior of a circle became famous motifs in his repertoire. Escher's works gained visibility among mathematicians during the 1954 International Congress in Amsterdam. Roger Penrose and Harold S. M. Coxeter visited a special Escher Exhibition at the Stedelijk Museum and both came away most impressed by his intuitive grasp of mathematical ideas. Soon thereafter, each began promoting certain aspects of Escher's ideas in mathematical form. It took another decade, though, for the name Escher to find its place in popular culture. A major breakthrough came in April 1966 when Martin Gardner featured drawings by Escher in his "Mathematical Games" column in *Scientific American*. The latter's name got another boost in 1979 with the publication of Douglas Hofstadter's Pulitzer Prize-winning book *Gödel, Escher, Bach*.

Another case of geometry intersecting with art took place in the mountains of Western North Carolina, where several displaced Germans gathered after Hitler came to power. Max Dehn, one of Hilbert's many gifted students, was a late arrival, but taught at Black Mountain College (BMC) from 1945 up until his death in 1952. As a multitalented geometer with strong aesthetic sensibilities, Dehn had the ability to adapt to the atmosphere at BMC, a liberal arts school with a strong focus on the arts. There he taught courses in philosophy and foreign languages, but also "mathematics for artists," a course in which students learned how to draw figures using principles from projective geometry, like the theorems of Pascal and Desargues. Dehn's course complemented the teaching of the influential art instructor, Josef Albers, who came to BMC with his wife, the weaver Anni Albers, soon after the college opened in 1933 (Menand 2021: 230–52). Both had been associated with the Bauhaus up until its closure that same year under pressure from Nazi government officials. Several who studied under Albers at BMC went on to become prominent artists, including Ruth Asawa, Ray Johnson, and Robert Rauschenberg. This phase in the college's brief history ended abruptly, however, when Josef and Anni Albers left in 1949; one year later, he became head of the Department of Graphic Design at Yale University. Figures like Man Ray, Escher, Albers, and Dehn underscore how crossover influences from European traditions profoundly influenced new developments within various contexts of American culture.

The interplay between mathematical forms and the visual arts has been varied and abundant, and yet this is only one of the aesthetic dimensions that plays a major role in mathematical activity. Some mathematicians express themselves most effectively on stage, and not a few may truly be called performing artists. Gustav Herglotz gave lectures that left his audience spellbound. Many who heard him thought he surely served as the model for his student, Emil Artin, who became a cult-like figure during his years in Princeton. In a review of one of the Bourbaki volumes, Artin stated his own personal credo as a mathematician:

> We all believe that mathematics is an art. The author of a book, the lecturer in a classroom tries to convey the structural beauty of mathematics to his readers, to his listeners. In this attempt, he must always fail. Mathematics is logical to be sure: each conclusion is drawn from previously derived statements. Yet the whole of it, the real piece of art, is not linear; worse than that, its perception should be instantaneous. We have all experienced on some rare occasion the feeling of elation in realizing that we have enabled our listeners to see at a moment's glance the whole architecture and all its ramifications. (Rowe 2018: 402)

Artin's purist views went far beyond Dehn's eclecticism. For the latter, mathematics stood between the humanities and the natural sciences, although its practitioners pursued a quite unique form of scientific method. Mathematical research was more spiritual than the natural sciences and more sentient than the humanities. Moreover, mathematical knowledge had vital links with the whole history of science, but also intimate connections with the history of philosophy. Finally, mathematicians represent not only a manifestation of *homo sapiens*, their practice reflects two other sides of human life: *homo faber* with an admixture of *homo ludens*.

CHAPTER ONE

Everyday Numeracy

What it Means in the Information Age

CHRISTOPHER J. PHILLIPS

In 1914, mathematics was one way among many of knowing about the world and was taught in schools largely as rudimentary reckoning and rote deductive reasoning. Just over a century later, the field has become *the* paradigmatic method of rigorous analysis. In schools around the world, success in mathematics has come to be seen not just as a general indicator of intelligence but also as a reliable path to employment.

This new importance for numeracy occurred through both intellectual as well as structural transformations. Intellectually, as science and technology moved from roles primarily in academia, the military, and industry to the forefront of many sectors in the twentieth century, numeracy was increasingly seen as a prerequisite to the study of many different fields. If calculating with weights, measures, and money were the key elements of everyday mathematics in the nineteenth century, by the end of the twentieth, the focus was on being able to function in a world of statistical averages, data trends, and probabilistic predictions. Professions from marketing to management, public policy to chemistry, had become disciplines with practices that required mathematical competence. At the same time, a level of numeracy became necessary simply to interpret a physician's advice or watch professional sports. A philosophical and scientific shift from a generally deterministic world to one governed by probability and chance was felt by everyday people. Though relatively few would ever understand the mechanisms behind formal statistical tools or the new technologies of machine learning, preparation in a wide range of fields increasingly included numeracy as a prerequisite. Numeracy became more than a basic ability to reckon with integers. It was a set of skills crucial for understanding the algorithms and uncertainties central to everyday life.

Structurally, numeracy acted as a gatekeeper or filter in a world that began to use education and "meritocracy" as engines for social mobility. Upward mobility became increasingly difficult for innumerate people. As mathematics courses moved from an optional to a central part of the curriculum in primary and secondary educational institutions, grade point averages, graduation requirements, and post-secondary school entrance examinations were predicated not just on computational ability, but often also on basic competence in algebra, geometry, and trigonometry. Even fields such as law, psychology, and medicine, which were not obviously mathematical in the nineteenth century, had effectively required numeracy as a prerequisite during the twentieth century, since students who could not master mathematics were seen as insufficiently talented for the work required. By the turn of the twenty-first century, in most parts of the world, access to many professions was cut off to those unable to demonstrate numeracy.

Even more profound than this change in the place of numeracy within social and economic structures was the way access to numerical skills expanded around the globe in this period. In 1914, only a tiny percentage of humans had access to even rudimentary mathematical education. Only a handful of countries provided their citizens any conception of free universal schooling, and beyond those places, mathematical education was reserved for the elite. The expansion of schooling and of mathematics' place in it meant that the subject increasingly was included as part of what *everyone* should know.

The very invention of the English term "numeracy" in 1959 was directly tied to this expansion, in that it drew attention to global disparities in levels of educational attainment. That is, the need to assess numeracy did not exist until the idea arose that mathematical competence was something most people should be expected to obtain. After the Second World War, international agencies began to collect and publish global data documenting numeracy. This had already begun with the creation of the International Bureau of Education in 1925, followed by the United Nations Educational, Scientific, and Cultural Organization (UNESCO). These systematic studies on mathematical attainment served to establish benchmarks and general conditions for educational expansion. Though disparities in access persisted (even as late as 1975, few countries outside of the largest economic powers educated girls and boys at similar rates, and even today rates are highly variable), the general expansion of access to formal schooling meant that nearly all children were exposed to basic numeracy education at some point. For many parts of the world, "basic" continued to mean arithmetic, and thus access to the benefits a broader understanding of numeracy offered remained uneven. Global statistics on numeracy and mathematical achievement are not as robust as those documenting literacy, but what evidence does exist suggests that both minimum standards and competency rates have increased slowly but steadily over time. Mathematics beyond basic arithmetic was no longer a specialized subject, limited to those training to become scientists, navigators, or traders; increasingly, ordinary people around the globe began to acquire some knowledge of mathematics beyond arithmetic. Only in the second half of the twentieth century, however, could the very notion of "numeracy" as a necessary component of education on par with literacy be conceivable.

SCHOOL MATHEMATICS: FROM CALCULATING TO REASONING

Among those lucky enough to have access to adequate schooling, there was a profound shift in the meaning of numeracy across the century. Mathematics education in the first decades of the twentieth century still largely meant basic reckoning—arithmetic—with a dash of algebra and geometry for advanced students. At the university level, trigonometric calculations, series, sequences, and calculus were predominant, with the primary focus of mastering enough mathematics to understand fundamental equations for various scientific fields. With the exception of a tiny percentage of students who would go on to learn technical mathematics at the university level, the ability to complete basic arithmetic tasks was essentially all that school mathematics formally required circa 1914.

In the context of a newly flourishing international mathematics research community, this relative lack of preparation at the school level was cause for concern for many mathematicians around the world. As both research mathematics and compulsory education expanded in the first half of the century, educators had to grapple with defining the ultimate purpose of mathematics courses. Many outside the realms of research mathematics saw the role of mathematics education in schools as purely practical. Consequently, they lamented requirements for graduation or admission to institutions of higher learning demonstrating competence in seemingly abstruse subjects like algebra and deductive geometry. On the other hand, for members of the newly formed International Commission on the Teaching of Mathematics (ICTM), founded in 1908 at the Rome meeting of the International Congress of Mathematicians, instruction in secondary schools should introduce students to mathematics for "its own sake," in the words of American mathematician David Eugene Smith (Roberts 2012).

The ICTM engaged many leading mathematicians, including Felix Klein (Germany) as President as well as Henri Fehr (Switzerland) and Charles-Ange Laisant (France) as co-founders of its affiliated journal *L'Ensignement*

FIGURE 1.1. A classroom scene from a mathematics lesson in 1948. Photo by Topical Press Agency / Getty Images.

Mathématique. It also had broad representation, with members from nearly all European countries as well as the US, Japan, Brazil, and Australia. The organization produced an influential series of publications, noting the relatively low global standards for mathematics curricula in the years after 1908. The issues were never simple: mathematics clearly encompassed *both* a crucial set of skills useful for a wide range of practical applications *and* an abstract collection of intertwined logically defined systems. Although international collaboration would largely cease between the two World Wars, the debate about the relative importance of teaching basic arithmetic or providing students with a somewhat deeper knowledge of mathematics—known colloquially as the "math wars"—would remain a constant theme that spilled over into the present century.

In practical terms, the opponents of rote arithmetic won the "math wars", as the acquisition of computational skills was largely de-emphasized at primary, secondary, and post-secondary educational institutions around the world. On one level, the reason for this appears obvious in retrospect: electronic handheld calculators emerged as affordable tools in the 1970s, providing a fast and reliable alternative to mental arithmetic. With calculator sales soon reaching well into the millions, by the 1980s nearly anyone in a wealthier country could perform computations electronically using the basic arithmetic functions on

FIGURE 1.2. The Little Professor, first released by Texas Instruments in June 1976, led to over a million sales by 1977. Designed as a testing device for children under 10, the device poses arithmetic problems after the user selects a difficulty level. Creative Commons.

these devices. Over the following years, in fact, cheap, but powerful calculators and electronic computer programs enabled students to manage even more advanced techniques, like finding roots, performing numerical integrations, or graphing complex functions. By the first decades of the twenty-first century, a student's mobile phone could function as a handheld calculator on par with some of the most powerful computational machines of the previous century. Arithmetic never disappeared from the curriculum, of course, just as it had not after the earlier invention of the slide rule. Many schools continued to emphasize that basic computational ability was important, if only to be able to use a calculating machine effectively. What did fundamentally change was the notion that one should teach arithmetic as an end in itself: the goal of mathematics class could not simply be the acquisition of ever more sophisticated computational techniques.

More broadly, the shift away from arithmetic was driven less by technological developments than by intellectual ones. Over the course of the twentieth century, scientists and mathematicians grew increasingly critical of the methods used in teaching their subjects in schools. These critics often claimed that mathematics textbooks over-emphasized computation, but worse still, their approach instilled *rote* computation. This critique was uneven globally, but most prominent in Western Europe, particularly France, Germany, Belgium, and Sweden, and in the mathematical communities which had close ties to them, including the United States, the USSR and the United Kingdom.

There were many constituent elements in the various mid-twentieth century efforts to reform the mathematics curriculum, but two factors characterize them above all: an emphasis on active rather than passive learning and an emphasis on unifying structures rather than specific computational skills. Neither was strictly speaking new. The idea of mathematical instruction based on active engagement rather than memorization had already surfaced in countries around the world from the first decades of the century. For example, the pedagogy of *Arbeitsunterricht* (learning by doing) emerged in Weimar Germany as an antidote to passive mathematical education. A few decades later, though in a very different vein in France, adherents of the influential Bourbaki movement insisted on reformulating all mathematics education from the ground up, emphasizing formal logic and the underlying structures of mathematics (see Chapter 3). By the early 1950s, there was a broad effort across much of the Western world to "modernize" the mathematics curriculum, represented by organizations such as the *Commission internationale pour l'étude et l'amélioration de l'enseignement mathématique* (CIEAEM). Science and mathematics must no longer to be taught as a fixed list of facts, these critics emphasized, but rather as a mode of inquiry, a form of reasoning in general. Learning mathematics as a collection of arithmetic facts was simply far too passive; students needed to understand how to *do* mathematics, and more importantly, *why* mathematics operated in the way it did. These "modernization" projects, often running under the moniker "new math,"[1] were uneven in execution and effect on students, but they were also highly visible and contentious among mathematicians and mathematics educators.

The reformers were not limited to mathematicians in Western Europe. Although South American educators were less active initially, Latin American countries soon acquired Spanish-language translations of newer American and

European textbooks. Exchange programs, often funded as part of Cold War initiatives, increased the dissemination of new mathematical research and corresponding pedagogical reforms to countries around the world. Even on the "other" side of the ideological divide, Soviet educational reforms in the 1960s also began to deemphasize calculation in favor of a unified notion of mathematics as logical reasoning, especially through far-reaching efforts aimed at promoting algorithmic thinking. In schools around the globe, mathematics classes cast aside traditional emphasis on rote memorization of arithmetic facts (Moon 1986). In this new setting, pupils learned how to reason in the "right" way—whatever that meant to those in charge of the reforms.

One of the most visible and consequential sites of school mathematics reform was the United States, partially because of its outsized role as a large, rich country in the 1950s with near universal education through age 16. Moreover, in the midst of Cold War competition with the USSR, the US federal government pushed millions of dollars into funding mathematicians and scientists to rewrite textbooks. To give a sense of what the shift looked like, we consider some of the most influential mid-century American "new math" textbooks: those produced in the 1950s and 1960s by the mathematicians and teachers involved with the publicly funded School Mathematics Study Group (SMSG). Led by Yale (later Stanford) mathematician Edward Begle, this organization had support from several prominent American mathematicians, including A. Adrian Albert (Chicago), Andrew Gleason (Harvard), Albert Meder, Jr. (Rutgers), Edwin Moise (Michigan/Harvard), Albert Tucker (Princeton), Marshall Stone (Chicago), and G. Baley Price (Kansas).

Whereas earlier textbooks had students build up skills by completing computation after computation, the new ones aimed at getting them to think more broadly about the foundations of arithmetic. Thus, instead of moving slowly but surely from 1 + 2 to 1 + 3 to 1 + 4, students would first learn that addition was actually the union of two disjoint sets, each of which could be counted, as could their union. Similarly, numerical order was a matter of comparing sets; by matching members of sets one-to-one, students would learn what greater and lesser "actually" meant. After all, numerals were just representations of numbers, which were just names given to equivalence relations between sets. Students were to think of integer arithmetic as one example of a set of objects that satisfies some general rules of operation. It was, so understood, merely a matter of convention that basic arithmetic operated in the usual way. With this more general level of understanding, students could learn to apply operations or arithmetic in different ways, using base 8 rather than base 10, for example, or exploring forms of modular arithmetic (11 + 2 might be 1 instead of 13, as in so-called clock arithmetic).

SMSG textbooks also emphasized that mathematics might not even involve figurate numbers at all. For example, students might learn the principles of

FIGURE 1.3. A traffic sign near the US border in Quebec reminds travelers from the United States that metric measurements are used in Canada. Photo by Owen Franken / Corbis / Getty Images.

arithmetic by way of a playing card: instead of ordinary addition and subtraction, the operations would now be rotating the card, flipping it over an axis, etc. The objects would be the various positions of the card, and just as with arithmetic, students had to identify the identity operation, inverses, and related properties. From first grade on, new textbooks tried to teach students that at the heart of mathematics were logical relationships (equivalent to, subset of, etc.) between abstract objects. Mathematics, in this conception, meant systematic reasoning, or carefully defining valid rules, operations, and logical relationships. It was more important to know *why* $2 + 2$ was 4 or 3^{-3} was 1/27 than to make those calculations quickly.

There were many critics of these reforms. Not all mathematicians were in favor of this "modern" understanding of their field, and even those who generally supported it sometimes doubted whether teachers had enough background to use the new textbooks adequately. Indeed, across many countries the reforms were eventually scaled back or adapted to better address the specific needs of teachers and students. The backlash was perhaps fiercest in the United States, where local school boards enacted educational decisions, whereas national reform efforts remained relatively weak. In the 1970s, schools across the United States witnessed a revival of memorization and rote calculation as part of the "back to basics" movement. The at times overheated rhetoric surrounding both the promotion and rejection of the new methods across the century can easily obscure our picture of how the mathematics curriculum clearly shifted during this period. Whereas schools around the globe continued to teach basic arithmetic calculations, they also increasingly trained students to recognize the structures underlying various mathematical concepts.

In many regards, emphasis on the school curriculum runs the risk of overlooking the ways in which doing "everyday" arithmetic (and cognition more broadly) have little to do with translating the content of the classroom into real life situations. When customers in grocery stores are deciding between various products, for example, they often have their own ways of calculating that bear only a vague resemblance to the formal modes of symbolic manipulation they learned in school (Lave 1988). No doubt, many older students eventually realize that textbook accounts illustrating the niceties of elaborate calculations very rarely give an accurate depiction of reasoning in practice.

Pedagogical movements emphasizing "constructivist" or "progressive" methods ebbed and waned in popularity. By the early twenty-first century, however, it was widely believed that numeracy required not just the ability to execute long division but also the ability to translate the messiness of lived experience into mathematical models and concepts in order to better understand the patterns of the past and make predictions about the future. This was no easy matter: many of the reformers quickly realized it was far easier to change textbooks or tweak government standards than to train thousands of teachers who actually did the work of introducing students to concepts. In much of the world, acute shortages of adequately trained teachers remained the norm. Even if mathematical thinking was increasingly portrayed as a general form of reasoning, such ability remained far from universal.

The curricular reforms of the twentieth century had an explicitly political dimension to them as well. In many cases, however, the politics involved proved to be crosscutting and complicated. The subject matter itself might be conservative or radical even as the pedagogical methods were progressive, or the methods could be traditional even as the content itself was novel. German word problems about percentages in the 1930s might deploy motifs alluding to the Nazi's agenda (e.g., "elimination of the Jews"). Yet, the same methods could be quickly refashioned into progressive textbooks for training students in other countries to reason flexibly and creatively.[2] After the Second World War, mathematics education was increasingly framed as a political initiative of the global military-industrial complex. Though science and technology had long been important in warfare, from medieval catapults to modern airplanes, the spectacular and seemingly decisive role of cryptography, radar, and the nuclear bomb in the war emphasized that success in the military realm required prowess in the scientific sphere. Training a new generation of youth mathematical skills ensured future technological and, contemporaries presumed, military primacy. The acquisition of numeracy was never solely about the mastery of skills; it was also always about the meaning and uses of mathematics more broadly.

On a smaller scale, in the midst of a global Cold War the mathematics class became a site in which larger debates about political and intellectual training took shape. These debates, in turn, had much to do with a certain tension that lies at the heart of mathematics, even elementary mathematics. On the one hand, as research mathematicians attest (see the Introduction), doing mathematics is a creative activity that involves exploring the properties of mathematical objects, which can lead to a recognition of surprising or unexpected relationships. On the other hand, mathematics is all about purely objective knowledge and logical rigor, the assurance of knowing, no matter who or where you are, that 2 + 2 could only be 4 (on the political significance of this apparently uncontentious fact, see Chapter 7). The pervasiveness of the Western stereotype of life behind the Iron Curtain was that of a closed-minded and autocratic society. For some educators, this picture meant that teaching mathematics as a set of rote skills risked suggesting to students that rigorous knowledge could not be creative or autonomous. That is, rote pedagogical methods risked portraying mathematics as undemocratic or antithetical to the ideals of liberal education in the "free world." As Howard Fehr, reformer and onetime head of the National Council of Teachers of Mathematics, explained, mathematics class—when properly conducted—should "liberate" students from the "shackles of authority" by encouraging them to discover intrinsic criteria for determining the correctness of answers rather than deferring to authority (Phillips 2015: 78). Mathematics may have human-chosen, even arbitrary, rules and structures, but those rules had logical consequences, whether they involved arithmetic in base 8 or the positions of playing cards. By emphasizing the logical structures underpinning mathematics, teachers could reframe numeracy to mean not just the ability to get the right answer but to reach that answer in the right way.

One upshot of the various educational reforms for the long history of numeracy was that more students gained exposure to more *kinds* of mathematics. Calculus and statistics were not a part of the pre-collegiate curriculum in 1914, but by the end of the century, many took these courses, even non-mathematically gifted pupils. With the widespread availability of electronic calculators and successive reforms pushing the importance of understanding the properties and principles of mathematical reasoning over and above the need to master arithmetic, numeracy came to mean a familiarity with mathematics as a mode of thought, as calculation done in context rather than to produce isolated facts. Even as mathematical knowledge came to be seen as something educated people possessed, so did mathematics change how we think *about* the value of training in general. Educators, in fact, have long employed mathematical measures were employed to evaluate the performance and efficiency of schools and their students. Two central examples of this were the standardized aptitude tests developed during the first half of the century and the grading of educational institutions that emerged during the second half.

Initially, the "mental test" became part of the standard arsenal of educational psychology in the years following the First World War. The dominant test in the US was the Binet-Simon Intelligence Scale, made famous by American psychologist Lewis Terman, but which actually had an international background. Terman developed it as a revision of work previously done in France by Alfred Binet and Théodore Simon, who in turn had been influenced by laboratory instruments developed by the German physiologist Wilhelm Wundt. These tests would become the intellectual forebears of some of the most influential twentieth century examinations, including the US Army Alpha Test, the Scholastic Aptitude Test (SAT), the Graduate Record Examination (GRE) and many others.

Although drawing on a long tradition of using a single examination to provide an overall ranking of intellectual prowess (going back at least to seventh-century civil service exams in China), the new testing regimes assumed that numerical ability, when tested alongside vocabulary and logic, might reveal an underlying native intelligence quite separate from any particular educational experience. One of the reasons mental, aptitude, and intelligence tests proved so successful was their suitability for a wide range of ultimate ends. Some early proponents were mainly interested in showing that intelligence was inheritable, thereby justifying existing social hierarchies and forced sterilization of the so-called feeble minded; others, like Binet, rejected any hereditarian interpretation and were concerned instead to identify children who needed special attention (Gould [1981] 1996). In the hands of Henry Chauncey, influential president of the US-based Educational Testing Service and advocate of the SAT, they were used primarily as a predictor of first-year grades at college. These tests sat at the heart of the rise of a meritocratic ideal, in which, for example, universities claimed to select among applicants on the basis not of the quality of their educational performance (privileging those with access to superior schools) but on the basis of a student's underlying "intelligence."

In the US, aptitude testing could be deployed as a double-edged sword, providing opportunities for intelligent but underprivileged students, while denying upward social mobility to those deemed irredeemably unintelligent. This was not a universal experience, however, as some countries, like France, with stronger traditions of influential educational institutions for a technocratic elite continued to utilize aptitude testing mainly as a way to identify weaknesses rather than as a sign of promise (Carson 2007). Numeracy was almost never the only requirement of these exams, which frequently tested vocabulary and reading skills. Nevertheless, as the subject seemingly most isolated from context and training, educators tended to regard prowess in arithmetic, algebra, and geometry as the best evidence of underlying intelligence. In large part, this was about the efficiency of sorting: people might disagree about which student was the best poet; there was far less room for dispute, however, when it came to judging mathematical talent. Test results demonstrating mathematical ability were required, in turn, for access to various fields of education or employment that had very little to do with mathematical prowess. Evidently, many regarded the latter ability as a general measure of intelligence applicable in many other fields or activities. By both requiring knowledge of mathematics to do well on such exams and then by expressing the results numerically, numbers took on a key social role. Institutions utilized intelligence testing not only to judge individual ability but also to act as a sieve within a competitive social and economic system based on meritocratic principles. Being numerate increasingly afforded opportunities that non-mathematical skills or abilities simply did not, while at the same time justifying access to those opportunities based on merit rather than wealth or social standing.

Moreover, as schools and universities found funding more limited in the years after 1970, testing came to be used not just for evaluating students but also for ranking the performance of educational institutions themselves. In the United States, for example, the various reforms associated in the early twenty-first century with "No Child Left Behind" legislation made the numerical analysis of standardized tests an essential ingredient in evaluating school quality, with potential consequences for its funding. In the United Kingdom, performance indicators and metrics conceived as part of the Research Excellence Framework (REF) continue to shape funding and investment in higher education. Though these two examples focus on different levels of the respective educational systems, there are evident parallels between these cases. Both required careful numerical assessment of achievement, quality, and productivity, and in each case, the ascendancy of economic notions of "accountability," "efficiency," and "return on investment" were decisive in shaping what constituted educational excellence. Numerical measures thus complemented technocratic and bureaucratic modes of management. It became impossible to talk about the quality of an educational institution without trying to quantify it.

Another visible aspect of this increasing use of numbers and data in everyday life arose in the realm of sport. While most citizens may not track REF ratings closely, many follow their favorite sports clubs with a keen eye, even if they are unaware of how data affect decision making. Sports statistics had long fascinated fans, but not until the last decades of the twentieth century did managers and owners begin to harness data in various sports in an effort to evaluate players and plot strategies precisely. In American baseball, the Oakland Athletics famously hired economists and statisticians to pore through performance data, hoping to find and sign underpaid players, i.e., undervalued relative to their production. In British football, Liverpool's club developed mathematical models of every pass, shot, and tackle to assess how well players and managers were performing—and which outcomes were likely attributable to skill and which to chance. Owners of clubs playing ice hockey, basketball, and many other sports have now hired statisticians, data analysts, and other technical employees to monitor player acquisition, management, and strategy. In a realm in which millions of dollars are at stake every day, it comes as no surprise that clubs have latched on to the latest ideas to gain an advantage. The term "moneyball"—initially popularized as the title of a book by Michael Lewis—soon spread to a variety of arenas. It generally signifies novel ways to replace tradition-bound practices: politics ("Moneyball for Government"), law ("Moneyball Sentencing"), education ("Is 'Moneyball' the Next Big Thing in Education?"), and criminal justice ("Lessons for Policing from Moneyball").

In this regard, everyday numeracy in sport follows the pattern of school mathematics: in 1914, sports statistics were no more than a matter of arithmetic. Fans summed totals and sometimes calculated averages. A century later, so-called advanced analytics required even casual fans to grapple with complicated, norm-adjusted statistical measures, which tried to assess the overall value of a player or a strategy with an elaborate mathematical model of performance. It was not merely a shift from qualitative to quantitative, but rather a shift in the kind of numeracy required. A conversation between fans in a pub, which once centered on the number of goals scored or games won might now involve measures of efficiency and expected values. Numeracy was inseparable from larger transformations in mathematics, from economics and game theory to operations research, even if that vocabulary was unlikely to surface during a debate in a pub.

As the sporting case suggests, schools and universities were just one of many kinds of institutions facing the "tyranny of metrics" over the twentieth century (Muller 2018). Everyday numeracy was required to make sense of the quotidian yet invasive ways numerical measures dominated our understanding of what worked. This growing awareness had relatively little to do with the activities of mathematicians or mathematics educators per se. It was rather a product of the rise of mathematics within the social sciences and humanities, particularly within economics, psychology, and sociology. When writing in the first decades of the century about the "new" methods available in educational research, psychologist Edward Thorndike famously set a research agenda by noting: "Whatever exists at all exists in some amount. To know it thoroughly involves knowing its quantity as well as its quality."[3] If someone wanted to know whether a school was good or a student was capable, it would soon no longer be adequate for an evaluator to interview students and faculty, assess educational and familial backgrounds, and consider the quality of pedagogical interactions. Thorndike's plea for precise measurement would certainly find adherents over the subsequent century. The quality of schools and pupils is now impossible to

FIGURE 1.4. This photo illustration shows a laptop signing into an internet betting webpage together with Euro bank notes on October 2, 2010, in Dinslaken, Germany. By that time, internet sports betting was such big business that prosecutors were actively investigating 250 suspects for attempting to influence the outcome of some 270 European football matches. Photo by Lars Baron / Bongarts / Getty Images.

talk about without numbers: test scores, rankings, admissions yields, graduation rates, etc. The meaning of numeracy for schools, and within schools, has fundamentally shifted over the long twentieth century.

NUMERACY IN DATA ANALYSIS

The use of intelligence tests and educational metrics was a small part of a much broader expansion of the collection of numerical data and the rise of sophisticated data analysis techniques in the twentieth century. Data, particularly numerical data, fundamentally changed what it meant to know complicated phenomena. In turn, numeracy expanded to include forms of data literacy essential to an everyday understanding of ourselves as well as our environments.

The collection and processing of numerical data to draw conclusions about the world and make predictions about its future was not new. Statistics and probability had origins closely intertwined with ideas about regularities. Two crucial examples in the early modern period were the collection of Bills of Mortality in London and the rise of aleatory contracts (that is, contracts concerning uncertain events) in Europe more broadly (see Chapters 5 and 6 of Volume 3). Contemporaries drew on the developing mathematics of games of chance to help shape ideas about fairness and expectation: calculating the wager a rational man would make in a game was no different than figuring out how much insurance to take out before a ship sailed. It may be too difficult to make reliable predictions about an individual, but by aggregating individuals, one can draw sound general conclusions in advance. You may not know if *this* ship will sink, or if *this* person will die of plague, but you can know ranges of numbers for how many might sink or how many people might die, and on that basis one could price insurance or prepare for future epidemics.

A major change that took place during the late nineteenth and early twentieth century was that individuals came to be increasingly identified by and associated with particular numbers. Insurance policies offer a good example. During the immediate post-Civil War years in the United States, "whole life" insurance policies were offered based on mortality tables: older applicants thus paid more than younger applicants, independent of their health status. Soon, however, innovators within the insurance companies showed that by taking measurements of the body and its condition, it became possible to predict more accurately the risk of an individual's death rather than simply relying on their age (Bouk 2015). By collecting biometric and personal data, companies began to claim the ability to predict more precisely how long *a particular person* would live. The effect was wide-ranging and laid the groundwork for the development of welfare programs in Europe and North America during the first half of the century. The administration of these new unemployment insurance and old age pension schemes depended directly on sophisticated predictions and pricing of individual risk: increasingly sophisticated data collection and numerical analysis was required to ensure governments would successfully meet financial obligations, which stretched out indefinitely into the future, even as individual lifespans continued to evolve in the present.

Similar transitions took place in many arenas throughout the twentieth century, as companies and governments began using ever more elaborate data collection techniques to make probabilistic determinations of an individual's future situation or behavior. For example, financial institutions traditionally advanced credit based on interpersonal knowledge. A banker was expected to know the applicant, and approving a loan meant vouching for the person's character—Was the applicant married? In good standing at local stores? Employed in a respectable job? Credit thus served as a proxy for moral character. The rise of numerical credit scores undermined that system. By the 1960s, the American company Fair Isaac (FICO) took a variety of such inputs from thousands of previous records, and used a mathematical technique called discriminant analysis to weigh the factors and produce a single number that would measure an applicant's credit "risk." (Similar companies exist in countries around the world.) That number was deemed to (accurately) measure a person's credit-worthiness based on numerical data reflecting their history of honoring debts and repaying loans in a timely fashion (Hyman 2012). This was part of a long expansion of systems of credit and debt, but it made explicit that a single number could replace a complex set of qualitative judgments about a person's credibility.

Similar transitions occurred in parole prediction systems, which used a variety of inputs starting in the 1920s to predict the likelihood current prisoners would commit another crime if released early. A century later, building on this basic model, programmers have created elaborate algorithms to assess the likelihood of a parolee's recidivism. In China, the government introduced a social control system in which a person receives a three-digit "social credit" score based on past behavior, which may restrict their ability to take out a loan or to buy a plane ticket. Not all of these new scores rely on the same technology or statistical methodology, but an underlying premise pervades many different realms of modern life.

What have these novel scores and data collection schemes meant for the concept of everyday numeracy? People around the world have started to be far more aware of the fact that their personal data was a valuable commodity, one that could have a major impact on a person's ability to move ahead in the modern world. Before borrowing money, or signing a lease, applicants increasingly had to present their numerical credentials, as it were. Accessing some of these typically required a national identity card (with its own unique numbers). A problematic credit score could severely limit one's possibilities, but soon entire industries emerged offering a potential borrower the chance to rehabilitate a low number.

Such ratings games moved in the other direction as well, with invitations issued to individual consumers asking them to "rank" their experience with companies or products. Crowd-sourced reviews on global e-commerce platforms like Amazon, Alibaba, Foursquare, or Uber can have an outsized financial impact. Individuals and companies increasingly realize that a single number associated with their name or brand might directly affect a range of possible futures, whether or not they have any direct control over the figure itself. Numbers were not only something that people used in transactions; modern societies increasingly came to picture people, products, and firms as themselves inherently numerical. Potential customers can hardly avoid having to decide if a 3-stars-out-of-5 rating should give pause when choosing a restaurant, or whether 3.9 versus 4.1 really makes a difference. Numbers, their implications and their meanings, have become unavoidable.

Reactions have not only been passive. In some cases, individuals have tried to push back against various uses of numbers, arguing against the deleterious effects of reliance on credit scores in housing or employment applications, for example, or protesting about the use of social credit metrics to blacklist political dissenters. In many countries the twenty-first century saw a growing backlash against using automated algorithms to make predictions about recidivism, risk, and fraud. In 2019, a special report issued by the UN warned about politically driven changes that took place under the banner of the seemingly innocent term "digital transformation," by noting: "Systems of social protection and assistance are increasingly driven by digital data and technologies that are used for diverse purposes, including to automate, predict, identify, surveil, detect, target and punish."[4]

Numbers have come to play an increasingly dominant role in the social and political lives of people around the world, even as rigorous definitions of "fairness," "justice," and "bias" remain inherently and inescapably contested. Critics of algorithms have explained that the notion of neutrality is a myth: facial recognition software requires preexisting "training sets" to learn; credit scores encode social inequalities (Benjamin 2019; O'Neil 2016). Crime statistics intertwine with historical policing practices, particularly if the data is collected in such a way as to justify or perpetuate existing racial and social hierarchies (Muhammed 2010). The problem, for many data skeptics, has little to do with whether some data or algorithms might be "better." It is a fear that tools such as statistical prediction software actually "bake in" inequalities under the banner of neutrality or objectivity.

Not all reactions have been negative. Some people have embraced the everyday production and collection of numbers about themselves. In the late twentieth and early twenty-first century, a self-tracking movement took off, as people started recording how many hours of sleep they enjoyed, or how many steps they took in a typical day. Although bathroom scales and other personal data collection devices had been around for more than a century—with the first public "penny scale" invented in Germany in 1885—by the early twenty-first century it was possible and even commonplace for people to generate data automatically on many biometric measures, heralding a "Quantified Self" movement. New "wearable" technologies of self-measurement promise a form of agency, with users able to take control (at least in theory) through tracking their own data. From counting steps and calories to tracking blood oxygen levels, enthusiasts believe they can productively mobilize their personal data. Whether seen as empowering or as dehumanizing, the increasing substitution of numerical data for lived experience has fundamentally transformed conceptions of individual subjectivity.

While many numbers represent how we think about ourselves as individuals, new forms of accounting have also transformed how we think about ourselves in the aggregate. The rise of sociology in the late nineteenth and early twentieth century brought forth studies of the interactions and interrelations of individuals in modern societies. Early sociologists, particularly Emile Durkheim in France, found the collection of numerical data essential for understanding such phenomena. In his influential study of suicide rates across Europe, Durkheim supported his conjecture by looking at suicide rates and averages across different groups. Later, scientists like Alfred Kinsey studied the formerly taboo topic of human sexuality through careful collection and analysis of data from individual surveys. In both cases, the findings expressed the characteristic social behavior of various groups in numerical terms. One could then inquire about whether some form of sexual behavior was common within certain groups or whether predominantly Catholic societies were more or less prone to suicide. The answers to such questions then depended on the aggregation of numerical data.

More popularly, opinion polling, formalized largely in the 1930s, created linkages between individuals not through neighborhoods or formal party affiliation but rather through similar responses to survey questions. Alongside private polling companies selling their services, governments also began to collect data through formal surveys and questionnaires, creating a way of measuring characteristics of a "public" (Igo 2007). There had long been ways of assessing public opinion, or at least the opinion of people who "counted," from newspaper articles to formal hearings. After the 1930s, however, it was possible for governments around the world to promote or reject policies based on the people's wishes, backed by the appearance of social science methodology. As with self-tracking, public opinion polls immediately had detractors, some of whom claimed that pollsters intentionally under-sampled certain portions of the population. From the failed prediction of the United States 1948 election to the uproar over the 2016 "Brexit" vote in the United Kingdom, there have been numerous cases when national polls failed to predict the "will" of the people. Nevertheless, everyone in politics recognized that the collection of data was crucial for assessing popular support.

The use of statistics as a mode of evaluating policy was hardly confined to public opinion polls. Increasingly after 1920, governments in North America and Europe turned to another form of accounting to aid in policymaking, namely cost-benefit analysis. This requirement put the onus on numbers to resolve thorny political disputes, as it meant measuring both costs and benefits prior to the deployment of new regulations, programs, or initiatives. Though eventually taken up by military strategists, environmental regulators, and especially economists, the original proponents were engineers, often in connection with water-works projects or other complex infrastructure programs (Porter 1986). Although certainly not evading politics—the calculations themselves always could be, and were, contested—the introduction of formal cost–benefit analysis as a standard tool of policy making in the twentieth century increased the importance of numeracy among government officials. The issues at stake were also often far murkier than those that arose in the business world. Public policy decisions involved weighing potential harm vs.

benefit in matters as varied as international trade agreements, safety measures in automobiles, or regulations on carbon emissions. In many cases, they required the pricing of seemingly priceless entities, particularly the value of human life or health.

In fact, the pricing of human life represents just one example of how entirely new *concepts* are created through measurement. A modern citizen has access to all sorts of numerical data about the world and its recent history, not just its population, but also global rates of poverty, educational attainment, and disease incidence. Some of these measures were the result of elaborate statistical recordkeeping undertaken by new international organizations like the League of Nations and, far more significantly, the United Nations. On a national level, concepts such as Gross Domestic Product and Unemployment Rate essentially did not exist before the 1920s. The very idea of a "national economy," formalized by mid-century, enabled countries to compare rates of growth or stagnation, as well as provide benchmarks to evaluate specific policy outcomes. The new government in China, for example, undertook an exhaustive statistical enumeration almost immediately after the Communists' triumph in 1949. By collecting "social facts"—a concept that harkened back to Durkheim's numerical approach a half-century earlier—leaders hoped to collect "new statistics for a new China" and show the progress made across the vast country under its new political regime (Ghosh 2020). Though many individuals continue to persist in describing an economy by drawing on their own personal experience, by the twentieth century it was possible to fit that experience into a much broader picture of how millions of other citizens were faring.

As was the case with metrics on individuals, supporters of social metrics, whether on a global or national scale, have often portrayed their approach to decision making as more progressive than older ones, since quantitative findings are arguably more "scientific" and less arbitrary. In many jurisdictions, laws require elected officials to justify new policies or regulations, which inevitably means presenting numerical data to support such changes. As with parole prediction algorithms and other "digital transformations," critics have claimed that many of the numerical measures simply hide bias and arbitrariness behind a veneer of statistical objectivity. As Muller's *Tyranny of Metrics* (2018) detailed, following so-called Goodhart's Law, when a measure becomes a goal, it ceases to be a good measure. In an effort to measure performance more rigorously and objectively—or so critics claimed—the introduction of metrics in many realms of life also brought about myriad attempts to "game" the system. Nevertheless, it has become impossible across much of the world to contest or challenge policymaking without recourse to numerical evidence: numeracy was essential to engaged citizenship.

The shift on a societal level was substantial: everyday numeracy might have once been about making change in a local store or surveying plots of land in a particular region, but in the twenty-first century being numerate means having the ability to think in terms of numbers *across* distances. Put differently, if one's knowledge of numbers and accounting was once local, personal, and subjective, it now had become increasingly embedded in a global, impersonal, and objective world. A classic example that illustrates this difference comes from the sale of commodities. In eighteenth and early nineteenth century Europe, traders generally measured, bought and sold grain in individual encounters without any broader need for standardization. In the global market of the twentieth century, on the other hand, these transactions have to meet the specifications of numerous larger institutions, like the European Union, North American Free Trade Agreement, and World Trade Organization. Labor statistics present another striking example. Unemployment may have looked very different—and was surely experienced very differently—in India, Germany, and Paraguay, yet drawing on labor statistics from all these countries, one could compare their

FIGURE 1.5. The display board at the Philippine Stock Exchange on March 1, 2007, a good day as the Philippine share prices closed 4.0 percent higher. Photo by ROMEO GACAD / AFP / Getty Images.

respective unemployment rates using a single measure. In the world of finance, strategists can look at Icelandic real estate, South African mining profits, and American soybean productivity and link them all through complex mathematical models of exchange and value.

For ordinary citizens, this has meant that everyday numeracy in an increasingly interconnected world was not simply about converting pounds to yuan or gallons to liters. It required thinking about the mathematical measures that were at the heart of the lived experience. Metrics that gauge global well-being, the popularity of policy proposals, or the success of an investment were omnipresent in fundamentally new ways. This transition certainly did not mean everyday Mauritians or Mexicans regularly engaged with or challenged United Nations' metrics in any sophisticated way. Nevertheless, people were increasingly aware that decisions at all levels were made on the basis of the numbers collected at home and abroad, even as they might note the gap between aggregate measures of well-being or wealth and one's personal experience. The previously unimaginable ability to compare experiences at an individual and societal level across the globe has become so commonplace and pervasive in the twenty-first century that people simply take this for granted. Few ever pause to consider the intellectual and physical infrastructure required in updating data on global oil prices, comparative happiness levels, or regional unemployment rates.

As data collection and numerical analysis became increasingly central to an understanding of the world, particularly in the age of electronic computers and sophisticated algorithms, people required a far broader understanding of numeracy to engage with the modern forms of life even at an elementary level. Numeracy here was not primarily about calculation, but about the way in which numbers could stand in for complex phenomena like public opinion, value, or creditworthiness. Knowing that the gross domestic product (GDP) did not include unpaid labor, such as that done by stay-at-home parents, played into debates about whether certain measures meaningfully represented the economy. To be visible, one needed to be counted, and to be numerate increasingly meant understanding what was counted and why. Understanding what numbers mean was essential to making everyday decisions about the value of goods and services for sale, candidates for elected office, or one's own personal health.

NUMERACY IN THE FACE OF UNCERTAINTY

Numeracy before the 1920s largely reflected people's experience living in a deterministic world. Arithmetic promised definite answers. While nineteenth century developments had complicated this picture, it was not until the advent of quantum mechanics that a fundamentally different conception of the world emerged. Since then the notion of indeterminacy has placed limits on strict causality, opening the way to a probabilistic view of small-scale physical phenomena (see Chapter 5). Many scientists and philosophers also began to think more expansively in the twentieth century about fundamental notions like association and inference, suggesting the need to reassess simplistic models of causal relationships. Though relatively few people concerned themselves with the niceties of particle physics or the philosophy of causal inference, everyday numeracy was widely understood in the context of the regularities, risks, and uncertainties that underlie so much of modern life. Nowhere was this more widely felt perhaps than in the realm of popular understanding of medicine and health. Modern patients need to be numerate to understand their own health status, let alone in making decisions about managing their health care.

Over the course of the twentieth century, medicine was a crucial arena for this shift in conceptions of numeracy in the context of uncertainty, not simply because of the near-inevitability of interaction with medicine across a person's life, but also because medical authorities increasingly understood health and disease in probabilistic terms. Though infectious and epidemic diseases still dominated medicine at the outset of the twentieth century, the rise of antibiotics, improvements in sanitation, and longer lifespans meant that a century later chronic diseases, non-infectious diseases like cancer, and diseases of old age were the more common experience of medicine across a growing swath of the world. At the same time, the role of detection and prevention—that is, minimizing risk—became an essential part of modern medicine. The idea of chance as a component of numerical methods in modern medicine became more prominent and visible in a number of different ways.

First, probabilities entered the picture because treatments only work on some people and they are only successful some of the time. Very few diseases accorded with the nineteenth-century model known as Koch's Postulates—named after the German researcher Robert Koch—which characterized etiology and therapy in terms of the presence (and removal) of foreign organisms. Instead, variability in response proved to be the norm. On its face, there was nothing new about the proposition that different people responded to the same treatment in different ways, as physicians were long familiar with variability in response to diseases. The mid-nineteenth century American physician Elisha Bartlett realized that statistical theory was useful as a tool for interpreting evidence from medical trials. This approach proved especially useful when evaluating the effectiveness of a treatment based on the outcomes in a large numbers of cases. Drawing on the data from these cases, medical researchers could ascertain probabilistically whether the differences in outcomes were likely to have been due to the treatment or merely reflected chance variation. Bartlett himself was trained in the so-called Paris school's "numerical method," as practiced by Louis-Denis-Jules Gavarret and Pierre-Charles-Alexandre Louis in the 1830s.

By the twentieth century, modern medicine had moved well beyond routine statistical tests. Researchers no longer merely assumed that meaningful differences in treatments were those that exceeded a certain "standard error." Rather, building on research done especially in agriculture, physicians developed a more rigorous approach to evaluating new therapies through careful design of clinical trials or observational studies (Marks 1997). In traditional medicine, physicians often took a philosophical stance to justify their therapies, an approach dating back to the ancient humoral understanding of health. Modern medical treatments, on the other hand, relied on empirical evidence and statistical tests to determine whether the results of a specific treatment were, in fact, "statistically significant." This meant, in effect, that no theoretical "explanation" was required. Beginning around the mid-century, physicians grew less inclined to speculate about causes and cures, as they became accustomed to relying on statistically significant evidence that a prescribed therapy yielded effective clinical outcomes. Moreover, thanks to data drawn from clinical cases, they could form a judgement as to whether one treatment was more effective than the next without any clear-cut understanding of the chemical or biological factors involved.

Second, the role of probability and statistics became even more visible in documenting how certain therapies or behaviors were associated with lower mortality or morbidity rates on average, particularly once medical practitioners began advocating for these primarily on the basis of numbers. Two of the most familiar examples involved the links found between cigarette smoking and lung cancer and those that tied the risk of heart disease to high cholesterol levels. Because no experiment could be conducted forcing individuals in randomly selected groups to smoke or to eat foods high in cholesterol, researchers had to design other types of studies in order to present evidence in support of the claimed association. A series of such studies, conducted in both Europe and North America, preceded the definitive statement by the US Surgeon General in 1964 that presented hard evidence linking smoking to lung cancer, as documented and published in the widely read pamphlet entitled "Smoking and Health." The Surgeon General's committee had found that, though there were associations with many other health effects, the only one that statistically rose to the level of "causation" was in the category of male smokers who had a higher risk of falling in the group of men with lung cancer than did those who did not smoke. (Later studies would extend this conclusion to women as well as to a wide range of other negative health effects from smoking.)

Similarly, though poor diet had long been associated with heart problems, solid evidence supporting this only first surfaced in the 1950s through a series of observational studies. Most famously, the Framingham Heart Study, which began in 1948, established a clear statistical correlation between specific elements of lifestyle and diet, especially serum cholesterol, and subsequent adverse cardiac events. A decade later, the Framingham Risk Score was being promoted around the world. This score encapsulated in a single number the likelihood a person would experience heart problems in the future; not long afterward, similar "risk score" calculators became available for a range of other health ailments as well. The familiar late twentieth century claim that a particular behavior, substance, or biomarker raises or lowers one's risk of disease by some precise amount has its roots in these mid-century study designs.

For some people, this foregrounded a controversial new role for governments and public health officials in regulating behavior. For everyone, however, it heralded a momentous change in our understanding of healthcare, as we know it today. Instead of thinking of health as a matter of feeling poorly and seeing a physician for advice—medicine as the alleviation of known symptoms—maintenance of health now required looking for signs of *future* ill health. Certain behaviors were now known risk factors, potentially inviting poor health outcomes, whereas exposure to certain substances and agents, from nicotine to pesticides, were seen as increasing the chance of disease. In a word, health literacy now required numeracy. Though even physicians struggled at times to articulate adequately the statistical issues at play, by the 1960s the medical world had come to realize the inextricable ties between the language of modern healthcare and the jargon of probability and statistics.

Third, the centrality of chance meant the experience of disease increasingly became one of surveillance. As effective cures for cancer continued to prove elusive, for example, other preventive strategies moved to the center. The medical community devised screening tests—mammograms, colonoscopies, Pap smears—and promoted these as essential components of clinical medicine. Many doctors and patients now had to interpret incidence rates and false positives and act accordingly. A redistribution of the burden in decision-making followed in the wake of the global patient advocacy movements of the 1970s. This resulted in more autonomy for patients in matters of health, but it also represented a shift toward a probabilistic notion of disease. That trend fed into what Robert Aronowitz called "risky medicine" (2015), a condition in which ordinary patients increasingly experienced disease not as binary (healthy or ill) but as a spectrum of numerical risk requiring constant surveillance and management. For example, a positive screening exam for breast cancer, regardless of subsequent testing or treatment, might result in a lifetime of enhanced surveillance and increased risk for the individual, even in the apparent absence of any disease. Across the twentieth century, medical experts have advocated one measurement after another as important for assessing overall health: weight, blood pressure, serum cholesterol, antibody levels, sperm count, blood sugar, etc. If individuals no longer conceive of themselves as simply "healthy," but rather as having a low probability of becoming ill in the near future, this suggests how essential numeracy has become for understanding our own well-being.

Medicine represents but one instantiation within a larger transformation toward what the German sociologist Ulrich Beck once called a "risk society" (Beck [1986] 1992). Though acknowledging that a lack of certainty has always characterized human knowledge, after the 1960s, Beck claims, there has been an explosion of "manufactured uncertainties," especially driven by scientific and technological developments. We now live in a world dominated by uncertainties of our own making, one in which many feel caught in a feed-back loop that constantly challenges their ability to keep pace. Little wonder such people may question whether our tools for managing or insuring against systemic risks are truly adequate.

People living in the early twenty-first century face a world of inherent and inescapable uncertainty. To live is to be at risk of contracting a disease, and attempts to lower those chances only result in new or different risk states, all requiring an understanding of probability. Some consequences are relatively straightforward—routine matters for all physicians and medical personnel who possess standard knowledge and statistical numeracy—but others are more subtle and require navigating overlapping claims of possible risks. Every decision to take a prophylactic drug or change a lifestyle carries with it implicit changes in the range of possible adverse health effects. Medicine is, of course, not unique in this respect. Similar trajectories in which risk and uncertainty have come to play a dominant role appear in discussions involving climate change, nuclear war, or population management. As many researchers have recognized, humans are not always successful at making sense of these risks, or at calculating risk in a way consistent with standard economic theory. Still, the risks are inescapable. Numeracy in this context has increasingly come to represent knowledge not just of concrete, deterministic situations, but probabilistic ones. Decision-making in many contexts today takes place against a background of inherently limited knowledge. In more and more spheres of human life, people encounter situations in their world that bear little resemblance to classical astronomy, with its predictable celestial events and deterministic orbits, but rather they appear far more like quantum events, where the uncertainty is profound.

Paradoxically, even promoting an understanding of statistical theory does not necessarily improve this situation. Because statistical claims are inherently probabilistic, critics can always argue that the lack of deterministic knowledge is fatal. Indeed, as Naomi Oreskes and Eric Conway have carefully documented, this line of argument often surfaces in scientific "disagreements" that are concocted and maintained for private gain (2010). They further note that so long as measures of statistical significance are the only allowable instruments for establishing correlations between data groups, many seemingly clear-cut causal relationships will remain open to dismissal as "unproven" because they fail to meet the conventional threshold of statistical significance required in order to support claims of causality. By the same argument, however, even mechanisms "proven" by this method can be dismissed as "uncertain" precisely because—by the very definition of statistical significance—there is always *some* probability, however small, that the inference was made erroneously. Numeracy in this context does not lead to mastery over the uncertainties of life, even if an understanding of statistics and probability might enable us to make informed decisions based on evaluation of possible risks. Rather, everyday numeracy requires an acknowledgment that nearly all rational decisions made in today's world—whether for or against a medical treatment, a particular public policy, or simply deciding among the more mundane options presented in life—involve weighing possible outcomes probabilistically.

CONCLUSION

Having described the profound shift in the meaning and importance of numeracy over the course of the twentieth century, it might be tempting to extrapolate these trends into the future, though it appears very far from clear that all of them will continue much further in the present century. What does seem beyond doubt, however, is that modern societies will continue to harvest numerical data relating to all aspects of everyday life. This being so, numeracy will surely remain essential for making sense of the metrics and scores that dominate our global understandings of value and meaning. Computational ability still has an important role to play, even though few any longer view this as an end in itself. For the vast majority of people, numeracy reflects a user's understanding of how numbers function rather than the skill needed to replicate the calculations behind those numbers.

That said, for the increasing percentage of people with access to higher education, mathematical competence has certainly become a good ticket for entering the job market at a high level and this will likely remain so in the future. Around 1900, biology and engineering were very different sorts of disciplines; that they are now often lumped into the complex of disciplines denoted by such acronyms as STEM (Science, Technology, Engineering, Math) or MINT (Mathematik, Informatik, Naturwissenschaften und Technik), shows just how transformative mathematical methods have become. Indeed, few domains of science and technology have remained untouched by them. Even literature, long the realm of subjective interpretation and close reading, increasingly relies on computational methods in studies of authorship, consistency of style, and rhetorical influence.

At the same time, the very success of machine learning and other modes of automation has also increased the likelihood that mathematical jobs may be at risk in the future. Indeed, the pattern set in the 1950s and 1960s, when electronic computers replaced human computers (see Chapter 6) may well recur again, potentially creating a world in which mathematics is more pervasive and yet less and less practiced. Tellingly, courses offered by mathematics departments represent an ever-shrinking percentage of the mathematics taught on university campuses. If the triumph of statistical and numerical methods paved the way for the automation of those methods by electronic devices, it would be ironic if the machines that embody that knowledge made comprehension of those very methods less common. This fear goes to the heart of debates about everyday numeracy in this period: Does an electronic handheld calculator render arithmetic ability superfluous? Does the use of statistical tests to adjudicate the outcome of experiments make it possible to have more knowledge with less understanding? Do prediction algorithms simply encode existing biases in such a way as to exacerbate inequalities by using a black-boxed tool? Do screening tests result in an elision between the experience of being at risk for disease and of actually having the disease?

Despite the various claims made for the benefits of a more mathematical education, increasing the percentage of numerate citizens around the globe has largely failed either to reduce political conflict or to resolve fundamental questions about fairness and equity. The fact that parole decisions, once made exclusively by a panel of fallible individuals, now might involve recidivism-risk calculators, or choices among prescription drugs, once made by deference to a single physician's expertise, might now involve statistically interpreted aggregated data from clinical trials, does not mean people are convinced they are living in a better world. Even with the expansion of everyday numeracy, many remained unpersuaded by the seemingly irrefutable hard evidence presented by a scientific community convinced of the value of vaccines or the risks posed by climate change.

It is perhaps fitting, then, that one of the clearest examples indicating that the expansion of numeracy need not lead to a consensus view about what the numbers mean has been the global response to the SARS-CoV-2 pandemic, which began in 2020. Though one would have expected that fields like virology and public health would take the lead in public discourse during a pandemic, it is telling that many people came to understand the disease and its threat largely through the language of statistical models and biometric data. The phrase "flatten the curve" became widespread in the early part of the pandemic, as governments encouraged drastic measures to reduce ("flatten") the number of cases below the medical system's capacity to handle them. In fact, the use of numbers to track epidemic outbreaks has a long history, and one of the crucial goals had always been to predict the future course of an outbreak based on present-day knowledge.

In 2020, mathematicians around the globe, from well-established groups to upstart epidemic modelers, used a variety of strategies and designs to take existing data on case numbers, hospitalizations, and deaths, in order to predict the course of the epidemic over the following weeks and months. These predictions, in turn, became essential for governments instituting various mitigation strategies, including business closures, stay-at-home orders, and mask-wearing requirements. As in the case of clinical medicine in the twentieth century, specific biomechanical evidence of how these interventions actually might work was initially missing or inconclusive: almost all justifications fell on the shoulders of the statistical modelers until effective vaccines could be developed.

In a world uncertain about a new and very alarming risk, it is perhaps not surprising that people turned to statistics. Still, the fact that doing so led to fierce debates about the proper response is highly significant. Even neighboring countries, like Sweden and Norway, adopted vastly different responses to the pandemic, based on different assumptions about the predictions of spread and vulnerability. Journalists queried scientists, pressing them to explain how much they actually knew, while politicians faced difficult questions about whether they were handling the outbreak better or worse than other political leaders.

Complicating any triumph of rational mathematical policies over the methods employed during the preceding century, there was no global reporting standard for what counted as a death due to the disease, let alone how to count cases or estimate total caseload. Within countries like the United States, with relatively weak national public health systems, different states adopted entirely different methods for reporting results. Even with robust national reporting standards, well-publicized mistakes occurred, as in 2020 when Public Health

FIGURE 1.6. Mexico's Hugo Lopez-Gatell showing a COVID-19 graph at a press conference on June 10, 2020 in Mexico City. This indicates that all 31 Mexican states and the capital were then below the "red level" for numbers of positive cases. Photo by Hector Vivas / Getty Images.

England misplaced 16,000 cases due to the use of outdated spreadsheet software. This, though, was only a minor incident compared with widespread problems and doubts about the reliability of official numbers from some countries. While numbers had clearly taken center stage in the response to the pandemic, this certainly did not mean that the agencies responsible for collecting, verifying, and analyzing that data had the means to do so. Even monitoring and recording seemingly simple data, like the number of COVID-related deaths, proved to be a fiasco on the global level. In 2022, the World Health Organization reported that the actual number of deaths worldwide due to COVID was around 15 million people, 9 million more than the estimate based on figures reported by national governments.

Even as people recognized the importance of numbers, they struggled to use them to provide meaningful answers. Questions swirled about the relative standing of countries in the fight against the virus, but how should one compare relevant statistics? Using raw numbers of fatalities, deaths relative to total population, or perhaps "case fatality" ratios? Few of the reporting agencies bothered to properly age-adjust pandemic data to take into account the relative age distributions of a country's population (a disease that affected elderly people at higher rates would presumably have a particularly devastating effect on countries with an older demographic profile). Examples like these could be continued *ad nauseum*; the point is not that one measure is the best but rather that, even if the data were accurate and precise, people are not always able to grasp what statistics has to say about important questions. It is nevertheless telling that compared with our knowledge about the last true global pandemic, the misleadingly named Spanish Flu of 1918–20, historians will have far more detailed information about the effects and trajectory of COVID-19. Numeracy has become far more essential over the intervening century, but that has not always meant increased understanding or consensus.

Numbers—and the numeracy required to make sense of them—occupy an essential cultural place in today's world. There is little doubt that this phenomenon has fundamentally changed the everyday experience of people around the globe, from how they understand their health and wellbeing to how they understand the quality of their schools, football clubs, or elected officials. The most optimistic claims of those pushing numbers into every aspect of modern life—suggesting that such a world would become more transparent, rational, and integrated—remain contentious and contested. Without doubt, however, everyday numeracy has expanded almost beyond recognition since 1914, both in the number of people for whom the term applies and in the very meaning of the term itself.

CHAPTER TWO

Practice and Profession

The Mathematical Sciences and Globalization

TOM ARCHIBALD AND DAVID E. ROWE

The present chapter, like several others in this volume, has been structured around three periods familiar from political histories. The vivid human drama that played out during the course of two World Wars touched the lives of mathematicians in numerous ways. This naturally sets off the period ending in 1945 from the post-Second World War era, dominated in political history by the Cold War conflict between the USA and the USSR. Throughout the period 1920 to 1933, international relations among mathematicians from the leading nations gradually improved, as highlighted by the 1932 International Congress held in Zurich. Four years later, at the ICM in Oslo, political tensions largely overshadowed the mathematical festivities (Hollings and Siegmund-Schultze 2020).

The year 1945 marked the starting point of a new economic world order, one in which mathematical models played a key role in both government and private economic policy and decision-making. Throughout much of the period up until 1990, competition between the two superpowers, the United States and the Soviet Union, created tensions across the international mathematical landscape. In research mathematics, the USA strengthened its dominant position, a process already begun before the Second World War but augmented by a large mathematical immigration to that country, which produced an explosion of educational and research developments during the postwar period. Meanwhile, the Soviet school continued to turn out brilliant researchers not only in pure mathematics, but also in scientific fields that depended on sophisticated mathematical methods. This period thus saw a vastly expanded use of quantitative techniques in the physical sciences, medicine, and engineering together with the widespread development and adoption of statistical methods as well as deterministic models to advance these fields.

The demise of the Soviet Union in 1990 hardly bears comparison with the collapse of Nazi Germany in 1945. Nevertheless, that year offers a convenient demarcation point for our present era since it coincides quite closely (and not coincidentally) with the revolution in communications associated with the rise of the internet. In 1990 Tim Berners-Lee, then working at the European nuclear research agency CERN, produced the first web browser and editor, later named WorldWideWeb. Together with the smart phone, the Web stands out as the most visible byproduct of the IT revolution with its roots in electronic computing, a field whose ties with the mathematical world have been far from incidental (see Chapter 6). Today, hardly any field of mathematical research has remained untouched by the fast-growing developments in computer science and technology.

MATHEMATICS AND INTERNATIONAL CONFLICT, 1914–1945

During the latter half of the long nineteenth century leading up to the First World War, forces of modernization reshaped life in Western Europe and by extension numerous other societies around the world when global imperialism dominated world politics. Mathematics, which forged ever stronger links with innovations in science and technology, played a major role in this accelerating modernization. Its practitioners shaped these forces just as they and their practices were molded by the various social, economic, and political contexts in which they lived. Mathematics and modernity were inextricably interwoven (Schappacher 2022: part II).

The ghastly period in European history framed by the two World Wars left a deep mark on the practices and professions associated with mathematics. International cooperation among scientists and educators had been steadily on the rise during the twenty years that preceded the outbreak of the Great War, though this had taken place alongside growing nationalism and militarism. European nations competed for power and prestige, extending their influence by means of the colonies they had acquired throughout Africa and Asia. These often served as outposts for scientific activity, as established by the British in India and South Africa.

Prior to 1914, several European mathematical societies had been founded and five international congresses had been held. The sixth, which Gösta Mittag-Leffler had offered to organize in Stockholm in 1916, did not take place, and the triumphant Entente powers initiated a general boycott against the defeated enemy nations that prevented virtually all international conferences for nearly ten years. The breakthrough for mathematics came when Salvatore Pincherle was appointed President of the International Mathematical Union (Lehto 1998). Since the IMU insisted that German mathematicians could not be invited to the 1928 ICM in Bologna, Pincherle got around this requirement by having his university serve as official sponsor for this event. While the international reconciliation was not complete, the IMU finally altered its policy and the 1932 congress in Zurich was held with something closer to full international participation.

The tense political atmosphere between the two wars greatly exacerbated the longstanding rivalry between France and Germany, enflaming nationalism on both sides: Émile Picard and Gabriel Koenigs promoting the French, L. E. J. Brouwer and Ludwig Bieberbach the German cause. Whereas Pincherle and his allies effectively broke the boycott at future ICMs, scientific internationalism remained much weaker than before. Competition for international influence—and for international students—also sharpened in this period. New alliances were difficult to form, and the leading internationalists—Paul Painlevé, David Hilbert, Mittag-Leffler, Arthur S. Eddington, Godfrey H. Hardy, and Harald Bohr—had to overcome many obstacles.

Italy's mathematical importance is underlined by the presence there of the Bologna congress in 1928 and by the successful efforts of Italian mathematicians to end the exclusion of the Germans and their allies. However, the rise of Fascism introduced schisms in the mathematical world, at first along the lines of social policy, and later, following the rise of Hitler in Germany and the formation of the German-Italian Axis, by a racial policy that saw leading figures dismissed from teaching and administrative positions. Vito Volterra, who had served in the First World War as Minister of War, and Tullio Levi-Civita, the famous architect of tensor analysis, were both edged out of all scientific positions.

Paris and Göttingen remained the two leading international centers up to 1933, closely followed by Rome. Italian analysts cultivated close relations with their colleagues in France, whereas several geometers remained on close terms with their German counterparts. In Göttingen, the explosion of interest in relativity and quantum theory (see Chapter 5) led to intense interactions between mathematicians and physicists, creating a natural network with Copenhagen (Niels Bohr) and Munich (Arnold Sommerfeld and his many pupils). Another kind of international network in number theory and analysis arose when Hardy introduced pure mathematics at Cambridge, creating links with Paris (Jacques Hadamard), Copenhagen (Harald Bohr), and Göttingen (Edmund Landau and Richard Courant). The 1920s also saw a burst of new activity in applied mathematics, long a British specialty. Three centers for research on fluid dynamics and/or aeronautics—Göttingen (Ludwig Prandtl),

Aachen (Theodor von Kármán), and Delft (Jan Burgers)—began holding international conferences in 1924. In 1921, Richard von Mises founded the *Zeitschrift für angewandte Mathematik und Mechanik* (ZAMM), which became the leading journal for engineering mathematics.

Mathematics Outside Western Europe

Significant developments in higher mathematics also took place outside Western Europe during this period (Parshall and Rice 2002). Up until 1914, most Americans who sought careers as researchers spent at least one year studying in Europe, usually at German universities. In the meantime, graduate programs at several leading universities in the US—Chicago, Harvard, Princeton, et al.—began to produce quite large numbers of PhDs in mathematics (Parshall 2022). Leading figures like George D. Birkhoff and Oswald Veblen followed developments across the ocean closely. Birkhoff traveled extensively throughout Europe as an advisor for the International Education Board of the Rockefeller Foundation (Siegmund-Schultze 2001). On his advice, the IEB granted funds for building the new Mathematics Institute in Göttingen as well as the Institut Henri Poincaré in Paris. The IEB also promoted the careers of a great number of younger mathematicians by means of research grants that enabled postdocs to work with senior European figures. Similarly, the Boxer Indemnity Scholarship Program, founded in 1908, enabled young Chinese students to study in the United States. One of its most prominent recipients, Chung Kai-lai, took his PhD at Princeton and became a leading authority on probability theory. Some returned to academic work in China: Hu Ming-Fu, the first PhD from this group in 1917 (at Harvard), became a professor in China, though the turbulence in China during the interwar period inhibited the flourishing of research there.

Already in the 1890s, Rikitarō Fujisawa founded a mathematical seminar based on the German model at the University of Tokyo. Fujisawa had spent five years studying in Europe, and brought back with him the zeal to train young researchers. His most talented pupil, Teiji Takagi, gained international acclaim for his work on class field theory. After studying in Tokyo, Takagi went to Göttingen to study number theory with Hilbert, the famous author of the *Zahlbericht*. He arrived only to discover that Hilbert had changed focus, but Takagi held fast to his original goal. This eventually led to a major breakthrough, though it was not until the mid-1920s, however, that Emil Artin and Carl Ludwig Siegel came to recognize the importance of Takagi's achievement (Yandell 2002). This was only the beginning for high-level Japanese research mathematics.

Higher mathematics also gained considerable momentum in several Eastern European countries. The collapse of Austria-Hungary and ensuing economic hardship led, however, to a serious brain drain after 1918 when a number of brilliant young Hungarians departed. Polish mathematicians, with the re-establishment of a Polish state, were finally in a situation in which instruction in Polish was permitted. Establishing research groups in several centers, they quickly established themselves as leaders in set theory, logic, topology, and functional analysis. The Polish journals *Fundamenta Mathematicae* and *Studia Mathematica* became leading venues for publications in these fields, drawing inspiration from Felix Hausdorff's 1914 monograph *Grundzüge der Mengenlehre* as well as from the Hilbert school. National research communities were also established elsewhere in Eastern Europe, notably in Czechoslovakia.

Soviet and American Mathematics

In Russia, the October Revolution and Civil War also led to large-scale emigration and loss of intellectual talent. Soon afterward, however, the Soviet Union stabilized and laid the groundwork for impressive achievements in many fields (see Chapter 3). Moscow emerged as the leading center for Soviet mathematics, led by Dmitri F. Egorov and Nikolai N. Luzin. Both, however, came under attack for their failure to promote socialist ideals in a period of shifting national priorities. In 1936, the Soviet Academy opened an investigation of the Luzin case, following which he lost all influence as his department at the Steklov Institute was closed. Nevertheless, several of Luzin's students, including Paul Alexandrov and Andrei Kolmogorov, soon emerged as leading figures in the Soviet school (Zdravkovska and Duren 1993).

Zjazd Kół Matematyczno-Fizycznych (Lwów 1930). 1 — L. Chwistek, 2 — S. Banach, 3 — S. Loria, 4 — K. Kuratowski, 5 — S. Kaczmarz, 6 — J. P. Schauder, 7 — M. Stark, 8 — K. Borsuk, 9 — E. Marczewski, 10 — S. Ulam, 11 — A. Zawadzki, 12 — E. Otto, 13 — W. Zonn, 14 — M. Puchalik, 15 — K. Szpunar

FIGURE 2.1. Meeting of Polish Mathematicians in Lwów, 1930. Wikimedia Commons.

An older tradition in Russian mathematics dates back to the career of P. L. Chebyshev, well known for his work in a variety of fields, including probability, statistics, mechanics, and number theory. Two of his most distinguished students at St. Petersburg University were Aleksandr M. Lyapunov, and Andrei A. Markov, who later assumed Chebyshev's chair as well as his position as a member of the St. Petersburg Academy. In the 1920s, three of Markov's best students emigrated from Russia: Abram Besicovitch became a fixture at Trinity College Cambridge, where he was appointed Rouse Ball Chair of Mathematics in 1950, whereas Jacob D. Tamarkin and James V. Uspensky found positions in the United States, the former at Brown and the latter at Stanford (Lorentz 2002).

These losses seriously weakened higher mathematics in Leningrad, a development that coincided with the rise of Moscow as the leading center in the Soviet Union. The former Imperial Academy of Sciences, originally founded in St. Petersburg by Peter the Great, was renamed the Academy of Sciences of the USSR in 1925, and in 1934 under Stalin its headquarters was moved to Moscow. Nevertheless, Leningrad continued to make important advances in physics and applied mathematics. A key figure behind this work was Yurii A. Krutkov, who worked alongside Aleksandr A. Friedmann at the Leningrad Meteorological Institute. Friedmann was the first to propose dynamical models for relativistic cosmology, and Krutkov helped to persuade Einstein of the merits of this view. Krutkov's student, George Gamow, became famous for his work on the big bang model for an expanding universe. Another pupil of Krutkov, Vladimir Fock, became a foremost authority on general relativity in the Soviet Union, at the same time promoting a Marxist interpretation of Einstein's theory. Applied mathematics and mechanics likewise had a long tradition in Russia, and in line with the government policy of working for the good of the state these areas were much emphasized in university education

The displacements that took place in Eastern Europe during the 1920s, however, were relatively minor when compared with those that began in 1933 when the National Socialists came to power in Germany (Siegmund-Schultze 2009). The Nazis quickly passed laws and decrees aimed at "cleansing" the civil service of all those who failed to satisfy the regime's standards for holding such positions. Those affected included academics with an unsuitable ethnic background, a provision aimed especially at persons of Jewish heritage. Since German Jews had come to occupy a larger percentage of university positions than before 1914, these measures led to an immense loss of intellectual capital for Germany. Many of the leading emigres from the 1930s found employment in the United States, the primary beneficiary of the Nazis' plan to purify the German nation as well as to dominate all of Europe. This emigration was intensified after 1938 due to the German Anschluss of Austria and the increased adherence of Italian fascism to the norms of Nazi racial policies. Two of the most famous emigres, Albert Einstein and Hermann Weyl, accepted positions at the newly founded Institute for Advanced Study (IAS) in Princeton. Those who could not find their way to the US were sometimes able to find refuge and employment elsewhere, for example in Russia, Turkey, and South America.

Research mathematics in the US flourished in the interwar years (Duren 1988; Parshall 2022). The IAS and the Princeton School of Mathematics were to occupy a special position in US and world mathematics, not only because of the power and prestige of the researchers there. In the early days, in addition to Einstein and Weyl, these included John von Neumann, Shizuo Kakutani, Paul Halmos, Kurt Gödel, and Wolfgang Pauli. Over the years, these names were joined by a stellar cast of both long-term members and eminent visitors. This conferred a special role on the *Annals of Mathematics*, a journal that had been located at Princeton since 1911. Setting standards for originality and what might be called good taste in mathematics, this journal promulgated the subjects of those publishing in it and did much to make them central. It remains in this leadership role, if not alone. During this period, many of the universities that continue as major research centers today first achieved that level. Joining the Ivy League schools and Chicago, schools in the Midwest (including Michigan, Indiana, Illinois-Urbana, and Wisconsin) developed strong departments, and in the West the University of California at Berkeley and Los Angeles, Stanford, and Caltech (the former Stroop Institute) became national and international leaders.

FIGURE 2.2. Princeton's Institute for Advanced Study. Wikimedia Commons.

Rise of Applied Mathematics in the USA

Before 1933, most American research had concentrated on a few areas of pure mathematics: group theory, algebra, logic, some areas in analysis, and foundations of geometry. Other fields—number theory, classical analysis, differential equations, mathematical physics—were barely represented at all. With the arrival of the Europeans, this situation changed dramatically, as applied mathematics gained a firm foothold through the efforts of Richard Courant and Roland G. D. Richardson. A native of Canada and longtime Secretary of the American Mathematical Society, Richardson became Dean of the Graduate School at Brown University in 1926. During the Second World War, he directed a center for applied research sponsored by the Navy's Office of Scientific Research and Development. This center employed a number of distinguished foreign-trained mathematicians, including the Russian, Tamarkin, the German, K. O. Friedrichs, the Austrian, Richard von Mises, the Czech, Charles Loewner, the Pole, Stefan Bergman, the Latvian, Lipman Bers, and the Yugoslav, Willy Feller—truly an international enterprise (Parshall 2022).

The other major center for applied mathematics, and ultimately for war-related research, was New York University (NYU), a second-rank department when Courant arrived there. Drawing on his many connections as former director of the Göttingen Institute, Courant quickly built up a strong program at NYU. Moreover, unlike the applied research center at Brown, which dissipated for a time after the war ended, Courant continued to build up his research group throughout the postwar era. Applied mathematics quickly spread to other institutions around the country. Another locale for war-related research was established at Columbia University in New York, which participated in the Manhattan Project as well as testing underwater sonar systems. At the IAS, John von Neumann made seminal contributions to the development and utilization of high-speed electronic computers. At the University of Michigan, Stephen Timoshenko, a Russian émigré, elevated engineering mathematics to a new level of sophistication, while Theodor von Kármán did the same for aerodynamics at Caltech. Von Kármán's wind tunnel at the Guggenheim Laboratory in Pasadena tested aircraft built on the West Coast during the 1930s. Richard von Mises, formerly the first occupant of a chair for applied mathematics in Berlin, took a similar position at Harvard. Stanford's George Dantzig developed the simplex method, a major tool in solving logistics problems via his technique of "linear programming," and MIT's Norbert Wiener created the new field of cybernetics, with implications for the theory of automatic control of machines and for computing (Heims 1980).

The career of Stephen Timoshenko illustrates how mathematical inclination and ability could afford enormous adaptability in a changing world. Culturally Ukrainian, Timoshenko received a mathematically heavy engineering education,

FIGURE 2.3. Courant Institute of Mathematical Sciences at New York University. Wikimedia Commons.

becoming an instructor in 1903 at the new Polytechnic Institute in St. Petersburg, where he became well versed in both British and German applied mathematical techniques. During his years there he began to visit Göttingen, where he was inspired by Felix Klein and worked with Ludwig Prandtl, a pioneer in experimental and theoretical fluid mechanics. Timoshenko was primarily interested in elasticity, and applications to topics like the buckling of loaded beams were to preoccupy him. He obtained a professorship in Kiev in strength of materials in 1906 and he attended the 1912 International Congress of Mathematicians in Cambridge, where he met many leading figures.

By 1921, when Ukraine was absorbed by the Soviet Union, he decided to emigrate, first teaching at Zagreb (in Russian, with an admixture of Croatian vocabulary) and then moving to the US to take up an engineering position in Philadelphia. In 1922, he joined the Westinghouse Research Group in Pittsburgh, where he helped set up a system for teaching advanced mathematics to company engineers, with general interest scientific lectures for the group and their wives on Sundays. Timoshenko moved back to the university world, first at Michigan, and after 1936 Stanford, retiring in 1944. While Timoshenko's wanderings, both institutional and international, were by no means unique, he was a pioneer in blending pure and applied research that proved fruitful in university, military and industrial settings. This pattern was to become much more common during the Cold War period.

Mathematics in Wartime

The watershed events that shattered the pre-1914 world order—spelling the end of the last European monarchies and the rise of fascist and communist governments in their place—brought at the same time technological advancements that created the social conditions we today identify with modern life. Among these innovations, some of the most significant have been those connected with modern warfare. Motorized air flight preceded the Great War, but state funding for research and development in aerodynamics increased manyfold as a result of its military significance (see Chapter 6 in Volume 5).

By the Second World War, control of the skies, which included terrorizing civilians living in cities, played a major role for both sides. Once the Allies gained the upper hand, the Germans hoped to recover with their new "wonder weapon," the V2 guided missile (*Vergeltungswaffe* 2, or "Retribution Weapon 2"). Designed by Wernher von Braun and his team, these were the first long-range missiles; during the last months of the war, the Germans fired some 3,000 of them on Allied cities, particularly London and Antwerp. Such technological know-how was of great interest to both the US and Russian military, as von Braun clearly realized. He and his research team gave themselves up to the Americans as the war was winding down; afterward the US military secretly brought over some

1,600 German scientists and employed them in various research projects. The Soviets later deported an even larger contingent of German scientists and engineers as Stalin's government took over control of Eastern Europe.

The entry of the US in the Second World War began with mobilization in early 1942, mathematicians being involved early on. Warren Weaver, who had served in the First World War, was a former department chair at Wisconsin and research director at the Rockefeller Foundation. He became director of the Applied Mathematics Panel of the Office of Scientific Research and Development. One significant achievement that came from this was the development of sequential analysis by Abraham Wald at Columbia; the purpose of this now widely-used set of statistical methods was to predict bomb hits from past data. A further interest of this office was computational projects, including the development of electronic computers (see Chapter 6).

The best known of the scientific projects during the war was the Manhattan project, which aimed to harness nuclear energy to build atomic bombs. A large number of European emigres took part in this work, with two mathematicians, von Neumann and Stanislaw Ulam, playing central research roles. Von Neumann, an expert on shaped charges, strongly advocated their use in constructing explosive lenses, which compress the plutonium core. This was the technology used in building Fat Man, the bomb dropped on Nagasaki at the end of the war. Ulam joined the secret laboratory in Los Alamos in late 1943 and was assigned to Edward Teller's group. Teller was keenly interested in developing a thermonuclear weapon, a project discouraged by the director, J. Robert Oppenheimer, who was skeptical that Teller's design for the "super" bomb could actually work. Soon after the war, Ulam and von Neumann were given the task of assessing this design, and they went about it with two different approaches. Von Neumann relied on electronic computers, including a twin copy of the MANIAC, which he worked with in Princeton. Ulam combined physical intuition with combinatorics and help from a team of female "computers," who carried out computations on mechanical calculators. Both sides arrived at the same result: Teller's design for the super was a dud.

The partnerships between mathematicians, physicists, and engineers that were developed over the course of the Manhattan project and its offshoots foreshadowed later developments, and the techniques for collaboration learned during the war were not forgotten in its aftermath. The interest in electronic computing for research in both pure and applied mathematics likewise carried on further, a development that enabled women to play an important role (see Chapter 6).

Women in Mathematics: a Cultural Struggle

Mention of female computers serves as a reminder of the peripheral status of women in mathematics, a situation that would only begin to change during the last decades of the century. Despite the examples of Sofia Kovalevskaya and Emmy Noether, the obstacles to female participation in the learned world persisted. These included lack of access to secondary education, confinement to social roles that prevented employment, and outright discrimination. Those few who managed to break through these barriers often found unexpected difficulties that kept them from realizing their aspirations (Green and LaDuke 2008, 2016).

In some ways typical for the United States during this period was the career of Mary Winston Newson, the first American woman to receive a PhD in mathematics from a European university (on her early years, see Chapter 2 in Volume 5). Although she had passed her doctoral exam *magna cum laude* with a dissertation approved by Felix Klein, when she returned to the US her prospects for employment were meager. She thus took a teaching job at the high school in St. Joseph, Missouri. Probably on the recommendation of Professor Henry Byron Newson from Kansas University, Winston attained a professorship at Kansas State Agricultural College. She remained in that position for three years, but then gave it up to marry Henry Newson in 1900.

Newson had studied at Ohio Wesleyan and Johns Hopkins, with periods in Heidelberg and Leipzig (where he attended a course taught by Sophus Lie). Due to nepotism regulations at the University of Kansas, Mary Winston Newson could only teach occasional summer courses. Over the next ten years, she had little time for mathematics. Not only was she busy raising the couple's three young children but she also had to care for her ailing mother-in-law. Nevertheless, she managed to find an hour or so each day to study, encouraged by both her husband and her mother. She also helped her husband prepare his *Theory of Collineations*, a book of 320 pages based on Lie's lectures from 1887–8. Sadly, Newson died from a sudden heart attack as this work was nearing completion, and he had no life insurance. Worse still, his death came six months before his widow qualified to receive payments from his retirement pension. As a final cruel stroke of bad luck, the English department at Kansas University had only recently hired her sister, the anti-nepotism rules thereby preventing Mary Newson from obtaining her husband's former position.

Three years later, Newson became an assistant professor at Washburn College in Topeka, Kansas. She remained there until 1921, when she became head of the department of mathematics at Eureka College, a small liberal arts school in Eureka, Illinois that admitted women. She taught courses in both mathematics and astronomy until compelled to retire from full-time teaching in 1937 because of her age, but continued teaching part time until 1942. Throughout her career, Newson took an avid interest in her profession as a member of the American Mathematical Society; she also served as President of the Kansas Association for Teachers of Mathematics. Yet, after her husband's death, she never had a chance to pursue mathematical research again. Such struggles were by no means limited to the US.

Brighter and Darker Sides of the Picture

It was during this period that the mathematical community began to devise international awards for research prowess, beginning with the Fields Medal, endowed by the testament of the Canadian mathematician John C. Fields. These medals, often called the Nobel Prize of Mathematics, in fact bear little resemblance to the Nobel Prize, in part due to their age requirement. Nobel laureates often receive this highly lucrative distinction when they are no longer active scientifically, whereas the Fields Award committee only recognizes young mathematicians who will be under 40 years of age at the time of conferral, which takes place just once every four years during a special ceremony held at the ICMs. The first awards, adjudicated by an international committee of the International Mathematical Union and awarded at Oslo in 1936, went to Jesse Douglas of the US and Lars Ahlfors of Finland. Emblematically, Ahlfors moved to Harvard as soon as possible after the war, only one of many in the worldwide emigration to the safe and rich USA.

We have alluded repeatedly to events, notably migration, that followed from the racial policies of the National Socialist government in Germany and its Italian ally in the period from 1933 to 1945. It is clearly impossible to convey in a few lines the destructive force exerted by these policies on a human level.

FIGURE 2.4 (a). The Fields Medals are conferred every four years during the ICMs. Wikimedia Commons.

FIGURE 2.4 (b). *Continued.*

The events that ensued underscore the fact that all mathematical activity is embedded in society, and it is often distressing to find that mathematicians whose work we admire were complicit in or supportive of these policies. These efforts at first deprived working and retired mathematicians in Germany of positions and pensions if they were defined as Jewish by the state. They included firings from editorial boards and the founding of a "truly German" mathematics journal, *Deutsche Mathematik*. Beginning in 1938 such policies were also enforced in Austria and in Italy. Most darkly, during the war the anti-Semitic campaign advanced to systematic murder. Reinhard Siegmund-Schultze has listed 145 German-speaking mathematicians who emigrated during the Nazi period to those places around the world which would have them; seventeen were murdered or driven to suicide, among them Felix Hausdorff (Siegmund-Schultze 2009). These figures do not include many others, for example those from Italy or Poland, such as Juliusz Schauder. Nor should one overlook the fact that many of the emigration stories did not have happy endings.

Racial discrimination was not present only in Germany, of course. To take a striking example, William S. Claytor, an African-American, was invited to the IAS in 1937 but could not take up the position, since he was banned from Princeton buildings on the grounds of his race. Taking up a fellowship at Michigan in the following year, he could not attend seminars for the same reason. Such discriminatory policies continue around the world in various forms.

MATHEMATICS IN THE SHADOWS OF THE COLD WAR, 1945–90

Before the war, political tensions had often played a detrimental role that affected international congresses in mathematics. During the 1936 ICM in Oslo, many representatives from Germany saw themselves as defenders of the new fascist state, whereas Mussolini's government refused to allow Italian mathematicians even to attend. Soviet mathematicians faced the same opposition from their government, although some had attended the two previous ICMs in Bologna and Zurich (Hollings and Siegmund-Schultze 2020).

After the war, such patterns continued, and mathematics remained involved in political matters due to a prestige that comes not only from its difficulty, but from its increasing social importance (Schappacher 2022: part III). There were no Soviet representatives at the 1950 ICM in Cambridge Massachusetts, by far the largest ever held to that date with 1700 participants. The 1928 Congress in Bologna had formerly held the record with roughly 800 attendees. After Stalin's death in 1953, the Soviet government continued its reluctance to allow travel abroad, even for Kolmogorov, a member of the selection committee for the Fields Medal at the 1950 ICM. When Grigory Margulis (b. 1946) was awarded the Fields Medal at the 1978 ICM in Helsinki, he was unable to attend because his government would not grant him an exit visa. Afterwards the Soviets proposed that their Academy of Sciences approve any Soviet candidates for a Fields Medal before they were conferred, a proposal that was immediately rejected.

Following the exodus of leading mathematicians from Central Europe and the devastation of the Continent brought about by the war, several leading French mathematicians took initiatives to restore harmonious relations with their counterparts on the other side of the Rhine. As one of the four occupying powers after the war, France had an interest in promoting cultural relations within the French zone. This included Freiburg University and the former Mathematics Research Center founded by its Rector, Wilhelm Süss, who headed the German Mathematical Society from 1938 to 1945. Despite the active role he had played in promoting Nazi policies, Süss could capitalize on the good relations he enjoyed with several French mathematicians. This enabled him to transform his research center into an international conference center, which flourishes today outside the town of Oberwolfach in the Black Forest.

Aside from a widespread willingness to forget the past horrors perpetrated by the Nazis and their Vichy collaborators, some French mathematicians continued to hold German mathematics in high regard. Several identified with the Göttingen tradition and felt inspired by the work of Hilbert, Emmy Noether, Emil Artin, and others. Those who first joined the Bourbaki movement, which began at the École Normale Supérieure in the mid-1930s (see Chapter 3), aimed to bring modern algebra, topology, and functional analysis into the mainstream literature. It would be difficult to exaggerate the influence of Bourbaki's structuralist legacy on European and American mathematics. The arrival in the United States of figures like Noether, Artin, Jean Dieudonné, André Weil, and Armand Borel deeply enriched the longstanding American research tradition in algebra and extended its more recent strength in modern analysis. Moreover, new waves of immigration from China, the Soviet Union, and elsewhere further diversified its expanding community.

Oscar Zariski (born in the Russian Empire, educated in Kiev and Rome) launched a flourishing school in algebraic geometry at Harvard, where his student, David Mumford, succeeded him. Marshall Stone built a new department at the University of Chicago beginning in 1946, with a stellar faculty. Within three years, Stone hired André Weil, Antoni Zygmund, Saunders Mac Lane, Paul Halmos, and Shiing-Shen Chern; the period was later referred to as the "Stone Age" (Duren 1989: vol. 2, 127–93). Halmos later moved to Indiana, whereas Chern went on to create an important center for differential geometry at the University of California, Berkeley. Zygmund, however, remained in Chicago where he established a school in harmonic analysis of enduring vitality. His students include some of the most influential analysts of the second half of the century.

Links between the newer French mathematics and activity in Germany had begun before the war, and were an important part of the rebuilding of mathematics in West Germany during the immediate postwar years. Friedrich Hirzebruch assumed a central position in this reconstruction, not only by virtue of central mathematical contributions, but also in connection with highly effective institution-building that was to reach across a divided Germany. These initatives strengthened international ties between Europe, the U. S. and Japan. Hirzebruch, called up in 1945 at age 17 into the German army but quickly captured, studied at Münster with Heinrich Behnke in the years immedately following the war. Behnke's editorship of the *Mathematische Annalen*, spanning the war years, made him both highly knowledgeable and well-known internationally. He recommended Hirzebruch to Heinz Hopf, who introduced the young man to the topological methods that led to his greatest contributions. From the point of view of mathematical institutions, however, the decisive event came with Hirzebruch's visit to the IAS in Princeton from 1952 to 1954, where his interactions with permanent faculty and brilliant visitors immersed him in a collaborative style of learning and research with international reach that he fought to reproduce on his return to Germany.

In 1956, Hirzebruch was appointed to a professorship in Bonn which soon emerged as the leading mathematical center in West Germany. There, he supervised the work of over 50 doctoral students during the course of his career. His initial effort at institutional innovation was the creation of the *Arbeitstagung*, an annual meeting of researchers with similar interests. The first of these, held in 1957, had only six participants, including Michael Atiyah and Alexander Grothendieck, who lectured for four half-days during the one-week meeting. These workshops eventually attracted around 250 people; anyone could be asked to lecture, on an assigned topic, with no more than two days' notice. For decades, this yearly international event became a fixture in the calendar of leading pure mathematicians (applied fields were favored in East Germany).

In the wake of the founding of the Institut des hautes études scientifiques (IHES) in 1958, a research center located in the suburbs of Paris and modeled on Princeton's IAS, Hirzebruch took part in two important institutional initiatives. The first led to the successful effort to secure long-term funding for the Mathematical Research Institute in Oberwolfach, whereas the second aimed to establish a Max Planck Institute for Mathematics in Bonn. The latter venture did not bear fruit initially, however, not least due to resistance from influential figures in the experimental sciences, an ongoing problem for mathematics internationally. Hirzebruch instead created a Special Research Center for Pure Mathematics (SFB 40) with federal funding, a test venture that proved successful and thus paved the way for founding the Max Planck Institute for Pure Mathematics in Bonn, which continues to attract high-level talent from around the world.

In France, the Bourbaki group dominated a large research community with fundamental activity in many areas, some of which dated back to the period before the war. The antiquated university system, however, fell under a great deal of social pressure already in the 1960s, a trend felt throughout Western Europe and North America; the "baby boom" did much to fuel this generational conflict. Student unrest, which had already been stoked by French wars of colonization in Vietnam and Algeria, now often aligned in revolutionary sympathy with Mao Zedong's China, Soviet Russia, or idealized images of them. This sparked street revolts in 1968, which soon led to sharp policial polarization within many academic communities, including the Sorbonne in Paris. These events had numerous and quite immediate consequences for mathematics, leading to demonstrations at the Nice International Congress in 1970 against the use of mathematics in war. Mathematicians, both young and old, took sides in these conflicts. One of the leading French figures on the left was the brilliant Alexander Grothendieck, while Stephen Smale, a Fields medalist, actively opposed US involvement in Vietnam.

Wider Impact of Social and Political Pressures

The "space race" and "arms race" put increasing pressure on educational systems to produce scientists and engineers. This was particularly true in the US, where the influx of European professors across the country served to highlight the often rather rudimentary character of technical education. In Western Europe, partly driven by socialist movements, Italy and the United Kingdom undertook programs to improve education in the public schools as part of a more general movement that sought to improve access to post-secondary education for all. Participation rates in higher education went up sharply both in the West as well as the Soviet Bloc; this also became a major priority, fostered from abroad, for governments in the "Third World."

Earlier political tensions continued to echo loudly on the international stage as well. The anti-communist furor in US society, which peaked around 1954, led to loyalty oaths and firings of some with left-wing backgrounds. Chandler Davis, a former student of Garrett Birkhoff at Harvard, left for Canada in 1962, and thus himself became an émigré transporting research mathematics to another country. Probably the most serious political impact on mathematics came with the Cultural Revolution in China, which led to widespread closures of institutions of higher education aand mass displacements of highly educated people to the countryside for re-education. The rebuilding and development of the Chinese universities that began in the late 1970s has continued up to the present.

American mathematics, as it emerged from the tumult of the Second World War, bore little resemblance to the tranquil interwar years. Not only had the cast of leading players changed dramatically, so had the institutional substructure that guided research efforts, which became more eclectic, reflecting the diverse roots of a highly internationalized mathematical community and the growth of government through the depression and war years. During the immediate postwar years, mathematicians of all specialties could supplement their university salaries with federal grants from the Office of Naval Research. This support expanded markedly after the establishment of the National Science Foundation in 1950. Universities no longer served as the sole benefactors for research mathematicians, as technology firms began hiring them for think tanks like Bell Laboratories, the RAND Corporation, or IBM's Watson Laboratory.

The link to technology firms was further cemented by the continual expansion in the development and use of the electronic computer, which gained considerable momentum from military interests. In 1952, the Atomic Energy Commission selected NYU as the first university to receive a high-speed computer, the UNIVAC. Courant hired Lipman Bers and Wilhelm Magnus, and he appointed Harold Grad to lead a group of researchers working on fusion energy and plasma physics. Two years later, NYU opened the Courant Institute of Mathematical Sciences, specializing in partial differential equations with applications to fluid mechanics and other fields. In the early 1950s, Cathleen Morawetz joined that program and later went on to become director of the Courant Institute.

New Disciplines and Professions

The fast-evolving discipline of computer science was frequently housed in either mathematics departments or engineering schools, as academic mathematicians interested in the use of such tools for both pure and applied research became more numerous. An early pioneer in this regard was Olga Taussky-Todd, who used contemporary computing tools on number-theoretic projects, both at the National Bureau of Standards (a federal government department) and later at CalTech. Military funding, notably from the navy, flowed to such projects and to think tanks like RAND. On the applied side, these interests dovetailed with more classical interests, for example in aerodynamics and fluids, that had preceded the war. These had direct impact on applied work, notably in the Bateman project to tabulate important mathematical functions, based at CalTech. They also were part of the pattern of growth of new application areas, such as operations research projects originally associated with military logistics. RAND was only one center for research in such fields. These efforts, as well as the mathematics of computation itself, gave a renewed emphasis to discrete mathematics. While we have emphasized here developments in the US, these had their counterparts in other large economies and in major military states. Outside the Soviet Union, the Eastern Bloc, and China, various private companies were engaged in such efforts.

Another large field of mathematical activity was statistical. Older forms of statistics, involving compilation of data, production of tables and the observation of trends in data, had long been supplemented by more detailed quantitative techniques, but the new grounding of probability theory on analysis that was formulated by Kolmogorov in 1933 spread rapidly. This meant that new techniques for statistical inference, like Ronald Fisher's analysis of variance from the immediate postwar period, could be put on a mathematically firm basis. Joseph Leo Doob achieved this in a 1934 paper entitled "Probability and Statistics," which gave a proof based on Kolmogorov's ideas of a central result in statistical inference. This likewise was followed by other innovations, like time series analysis, invented by Harold Hotelling, Jerzy Neyman, Abraham Wald and others during the Second World War. The burgeoning applications of these techniques and others required a high degree of specialization, and already by the 1960s departments of statistics began to appear in US universities. Agricultural applications, which had originally interested Fisher, led to expanded interest in statistical approaches in the biological and medical sciences. By the mid-sixties, there were specialized Departments of Biostatistics in North America, with emphasis on epidemiology. Drug trials, both in university research and by drug companies, depend heavily on statistical verifications in testing effectiveness, identifying side effects, and gaining regulatory approval. This is a pioneering example of one of the more important roles of mathematics in contemporary society: that of providing concrete quantitative criteria suitable for use by government regulatory agencies. Older techniques of descriptive statistics and projection of numerical trends remained popular with European governments, notably for demographic work, a central tool of domestic and colonial policy. This was particularly the case in France.

Early in the postwar period the applied mathematical and statistical problems that were accessible to successful solution were often highly idealized, because the computers of the early period, up to the 1970s, were slow compared to those today. Developments in hardware, in processing strategy (the use of parallel processors, for example) and in software all contributed to an exponential increase of computing capacity, one which was matched by the development of new methods for producing approximate solutions. The mathematical side of this has to do with specific methods that can take advantage of what speed there is. An emblematic early development was the FFT, or Fast Fourier Transform, a decisive step in reducing the computational complexity of routine operations. This was devised specifically in the context of a national security concern in the US, namely the detection of Soviet nuclear tests in the 1960s. The Finite Element Method in the computation of solutions for partial differential equations likewise came into its own in the 1970s, as a result of research that was widely distributed among scholars in the West.

In part because of this "boom," resources for mathematical research and teaching continued to expand throughout the fifties and sixties, but by the late 1970s the boom had turned to a bust. Caught between a sharp downturn in employment possibilities for pure mathematicians within the academic arena and a growing need for computer scientists and applied mathematicians outside it, the American mathematical community braced itself, hoping to cope with the demands of a tough new economic environment in which technical knowledge had become increasingly critical. Despite these challenges, the US community continued to grow and remained the leading driver of innovation in mathematics, though over the period up until 1980 a substantial recovery in the Western European community took place and international communication generally improved. The increasing use of English as a lingua franca was well established by about 1980, by which time the sheer proliferation of published research was becoming ever more difficult to catalogue and review. New fields seemed to be popping up everywhere, making the replication of a project like the Japanese *Encyclopedic Dictionary of Mathematics*, which in 1968 attempted to survey all current research fields in pure mathematics, highly impractical. At best, one could hope to gain an overview of the main activity through collections of essays, such as Pier (2000).

The role of women in the sciences varied with the developments just described. During and immediately after the war, women participated more and more in various forms of scientific research, usually in supporting roles. One result was that they began to enter programs in the mathematical sciences in increasing numbers, and to continue to graduate study. The various waves of the feminist movement challenged traditional male roles and the view that "math is not for girls." Despite activist efforts, however, and despite strong records of achievement, the participation rate of women in the research world, and in the mathematical sciences generally, has been slow to approach that of the male population. Whereas the total number of new PhDs in the US increased dramatically during the 1950s and 1960s, the percentage of these degrees that went to women dropped to around 5 percent, compared with a high of nearly 15 percent during the 1930s. During the 1970s, however, over 1,000 doctorates in mathematics were awarded to women, who comprised 10 percent of the total, whereas during the 1980s over 15 percent of these degrees went to female candidates (Beery et al. 2017: 94). Despite the barriers presented by institutional obstacles to female participation and success, institutionalized belief systems that limit women's roles, and sexism in all its forms, women participate more than ever worldwide in all aspects of the mathematical sciences, including the professoriate.

MIGRATION, DIVERSIFICATION, INTERNATIONALIZATION, 1990–2020

Following the fall of the Berlin Wall and the reunification of Germany in 1990 came the collapse of the entire Soviet bloc and, with it, the end of the Cold War era. In the meantime, former hostilities between the United States and China had also also eased, leading to new trade agreements and increased intellectual mobility. For a brief time, it seemed that the new world order would eventually follow the lead of an increasingly integrated system led by liberal democratic nations following the American model.

Like nearly all facets of post-industrial societies, mathematics has globalized and diversified in recent decades, while retaining a highly stratified professional structure. As a measure of elite status in the world of mathematics, Fields medalists reflect the continued dominance of a small number of institutions in a select few nations. Of the some 70 medals awarded since 1936, more than half of the recipients came from just five countries (US, France, Russia, UK, and Germany). These numbers, however, only tell part of the story because during the period from 1990 to 2020 nearly all the medalists, no matter where they came from originally, held affiliations with only a handful of elite institutions in just two countries: the United States and France. Russian mathematicians have reaped eight Fields Medals, beginning with Sergei Novikov in 1970. Yet, nearly all the winners from the post-Soviet era later took positions outside Russia: in Israel, the US, or Western Europe. For example, one winner from 2002, Maxim Kontsevich, even completed his doctorate outside Russia (in Bonn, Germany, with an American supervisor, Don Zagier) and now lives in France. This phenomenon reflects, on the one hand, the strong pull of elite centers in the West, which draw strength from their international connections, not to mention their salaries. On the other hand, there was a strong push factor exemplified by the many Soviet Jews who left Russia after 1990 to take positions in the US or Israel. Indeed, the draw of the US and various Western countries has been strong since 1990, both for those working in research and for the growing number of highly-trained mathematicians who work in business and industry. Increasingly, high-tech employers requiring mathematical proficiency take the place of universities, drawing well-trained creative mathematicians from around the world to firms like Alphabet and Amazon.

FIGURE 2.5. Honorees at the opening ceremony of the International Congress of Mathematicians held in Hyderabad, India, on August 19, 2010. Pictured are the four Fields Medals Elon Lindenstrauss (*left*), Ngo Bau Chau (*2nd left*), Stanislav Smirnov (*3rd left*), Cédric Villani (3R), 2010 Rolf Nevanlinna Prize winner Daniel Spielman (*2nd right*), 2010 Gauss Prize winner Yves Meyer (*right*), and Chern Medalist Louis Nirenberg (*4th left*). Photo by Noah SEELAM / AFP / Getty Images.

In the period since 1990 mathematics has continued as an extremely vital academic discipline; the increasing demand for computer scientists, and the increasingly quantitative features of almost every area of research in science and engineering, have continued to increase the demand for instruction and competence at all levels of post-secondary education. There has also been a tremendous diversification in the use of sophisticated mathematical tools in a variety of non-traditional areas. Outside of statistics and computer science, probably the best known of these has been in the mathematics of finance. That area depends heavily on stochastic partial differential equations, a subject that requires training beyond the undergraduate level; these tools now dominate the pricing of large classes of investment instruments. Another area in which mathematical tools have become fundamental is medicine; here it is sufficient to mention imaging, the CT scan and MRI, which have become indispensable tools for diagnosis in the past 30 years. This is one of the many areas in which dealing with large data sets has sparked mathematical research: can we get images faster, with less data and computation, that provide the same or better information? This is useful in any field that models using data: climate modeling being one of currently critical concern. Here, once again, the fruits of mathematical research guide policy, whether commercial or governmental. But probably the most notable growth in mathematical employment has occurred in the general area of security, and in particular cybersecurity and the detection of crimes perpetrated and managed over the internet. These areas too are vital both for commerce and for government.

The pervasive use of mathematical tools in all these areas means that mathematicians are found in every corner of business and industry where quantitative tools hold sway. Their situation in some ways resembles that of the late eighteenth century; not only must today's mathematicians be versatile and prepared to master new techniques, or adapt them for use in a new area; they also find their employment subject to political and economic vagaries to a much greater extent than was the case around 1900. At the same time, academic, or purely academic, mathematicians are coming to be outnumbered by those in industry and government, even in a context where we speak of those primarily engaged in innovation. More mathematical discoveries are now proprietary objects, guarded as secret, than was the case a century ago. Thus, while pure mathematics continues to flourish, academia now focusses increasingly on applications in this broad range of new areas as well. PhDs often train specifically in such areas and move to positions offered by companies or agencies.

Reflecting these developments, institutions that support mathematical research are likewise diversifying. What might be termed the "Oberwolfach model" of one- or two-week retreats for groups of researchers has greatly expanded, a large number of countries now have one or more such centers, and applied mathematics of all kinds is increasingly represented in the subject lists

FIGURE 2.6. Oberwolfach Workshop on Geometry and Physics of Higgs Bundles, 2019. Oberwolfach Photo Collection.

there. National strategies for providing funding to mathematical research vary widely, but there is increasing availability of funds for applied research, especially in the context of industrial partnerships.

Mathematics Education and Research

Because of all these developments, we see a new emphasis on STEM subjects in primary, secondary, and higher education (see Chapter 1). However, access to these subjects is largely governed by mathematical training, thought of as a filter for intellectual ability as well as a necessary tool. Access to such training thus becomes a precondition for participation in the economy at any level. Accompanying these developments were shifts in educational content. By 1990, while there was still a shared suite of basic (undergraduate-level) knowledge, the mathematical sciences (which include statistics, operations research, and much computing science as well as the more traditional fields) have become so diverse that providing a broad survey education that introduces students to most areas is now impossible. Specialization is such that most research papers can be read (and are read) by only a few people.

The total amount of research production is now vast. The internet, a development with early roots that came into its own in the 1990s, makes published research accessible, though there are often financial barriers to access. Thus, the shift to electronic publishing was not accompanied by a corresponding shift toward free access that public libraries traditionally provided. Unpublished research is also shared in various communities, for example via ArXiv, an internet site where research can be posted, or among research groups, whether private or public. Such shared research is greatly abetted by the mathematical typesetting tool TeX, originated by Donald Knuth in the 1970s, but which became universal after the invention of the internet.

Despite this technical accessibility, the world of mathematics is not remotely surveyable due to its size, and identifying trends in traditional ways is an elusive goal. One can learn, using the abstracting services MathSciNet or ZbMath, about the existence and content of literally millions of articles and books: MathSciNet, for example, contains 3.6 million items in its database of mathematical research, adding about 125,000 annually at this time, and reviewing about three-quarters of them. This compares to 2800 papers from the year 1940; at that time it was possible for a mathematician to scan all the titles and read the abstracts of all those he or she found to be remotely interesting. Clearly, there will never again be a program for future research like the Hilbert problems.

And, speaking of surveyability, significant changes have taken place in what consitutes proof, with "computer proofs" of various flavors, or "probabilistically checkable" proofs providing what many regard as sufficient certainty (see Chapter 6). If the scale of mathematics changes what is thought of as proof—going beyond the kind of debates about proof that have been associated with incomplete theories such as string theory—this certainly also changes what one thinks of as basic knowledge, and ideas about this have become much more diverse than was the case even a decade ago.

Training at the predoctoral level frequently aims at specific careers in business or industry, often (in Europe and North America) with Bachelor's or Master's-level programs that include internships and train for specific tasks. Actuarial science is probably the most highly developed form of this, with a sequence of examinations controlling entry to a lucrative and secure profession. Training at the doctoral and postdoctoral levels increasingly seeks to accelerate access to the frontiers of a specialty. Thus, the kinds of debates that took place in the twentieth century among practitioners concerning the nature of mathematical knowledge no longer occur except as cocktail party conversation, with few exceptions. And increasingly, given the requirements for specialization in mathematical careers, there are debates and confusions about what constitutes basic knowledge, where this belongs in university training, and how to equip researchers for the future. All of these trends have an impact on practice and profession in every corner of the complex mathematical world.

In part because of its importance in such fundamental areas as defense, security, technological design and determining the significance of experimental results, facility in mathematics has emerged as a priority everywhere. While this has been the case for some time in what used to be known as the "developed" world, today it is true around the globe and in all socioeconomic circumstances. The biggest single transformation of the past three decades has taken place in China. With almost a fifth of the world's population, China floundered in social chaos and profound economic misery prior to the revolution in 1949. For decades thereafter, internal issues occupied the Chinese government, and the country's contacts with the external world, including the mathematical world, were minimal. Since around 2000, however, the opening of China to trade, the renewed mobility of the population to undertake foreign training, and the largest and fastest economic development the world has ever seen, all have led to China becoming a mathematical world power, with the explicit goal of becoming the world leader. Whereas in 1950 China's universities were few, there are now over 2,500 post-secondary degree-granting institutions; almost all of these teach some form of mathematics. In China and elsewhere, such studies, especially at an advanced level, increasingly require literacy in English. Study at English-language institutions is much prized, not only in China.

Measures of Mathematical Excellence

Some institutional characteristics that appeared in the nineteenth and twentieth centuries continue to persist, though sometimes in new forms. Professional mathematicians who are university professors and researchers still often belong to national associations, like the AMS, SIAM, SMF, DMV, but also statistical societies, etc., and attend large-scale conferences for the diffusion of knowledge. These meetings are also increasingly proving grounds for graduate students. Merit in research likewise continues to be rewarded by prizes; these have become more diverse, encompassing work specifically in various areas of application. The Fields Medals remain important, conferring influence and affecting research trends. The influence of cash prizes, like the Clay Foundation Prizes for certain classical problems like the Riemann Hypothesis is harder to assess. Merit is also now routinely "measured" using various quantitative and non-quantitative indices. In the world of private industry, the primary metric for success is measured in the profit made for the firm. In the university world, it can be identified in various ways: sometimes by the degree of difficulty for publication in a given journal, at other times via impact factors, h-indices, and other scientometric tools.

In recent years prizes recognizing mathematical excellence have multiplied, and include the recognition of lifetime achievement. The Abel prize is the most prestigious of these, with the first winner in 2003 being the Bourbaki member Jean-Pierre Serre. The Western Europeans initially dominated the ICM meetings, but during the post-war era, other countries began to host the event, including the United States (1950, 1986) and the Soviet Union (1966). In 1990, Japan became the first Asian nation to host an ICM, held that year in Kyoto. China staged the 2002 ICM in Beijing, India hosted the 2010 event in

Hyderabad, followed by ICM 2014 in Seoul. In 2018, Brazil became the first South American country to stage an ICM, which took place in Rio de Janeiro. Reflecting the continued political context of mathematics, the ICM for 2022, slated for St. Petersburg, had to be moved online in response to the Russian invastion of Ukraine. The first woman to be awarded the Fields Medal was Maryam Mirzakhani, an Irani national working in the US, who received this high honor in 2014; the American Karen Uhlenbeck won the Abel Prize in 2019.

Who can become a mathematician now? The diversity of these communities is enormous. The still-improving access to higher education worldwide makes the possibility of acquiring mathematical training greater than at any time in history. The rigors of the subject mean that the classical remark attributed to Euclid remains valid: there is no royal road to mathematics. But there are some new roads and for more people than ever before. The resulting set of professional and occupational categories and practices imbues society with mathematics to a greater extent than ever before.

CHAPTER THREE

Inventing Mathematics

The Soviet School and Bourbaki as Two Visions of Modernism

ROGER COOKE AND LEO CORRY

Over the course of the twentieth century, mathematical invention was profoundly influenced by two somewhat idiosyncratic, but also quite distinct research programs that arose in Eastern and Western Europe, respectively. The first was associated with the Soviet school of mathematics, which took shape between 1917 and 1945. Its influence continued throughout the century and extended well beyond the borders of the Soviet Union to nearby countries, like Poland and Hungary, but also beyond to the United States and Israel. The second research program, developed by a far smaller group of mathematicians that first formed in France during the mid-1930s, later became famous through its collective works, *Eléments de mathématique*, published under the pseudonym Nicolas Bourbaki. Many of the group's original members were graduates of the *Ecole normale supérieure* (ENS), where the Séminaire Bourbaki was founded after the Second World War. As a movement that aimed to establish new foundational principles for modern mathematics, Bourbaki's impact gradually spread throughout Europe, the United States, and Latin America.

These two programs led to large-scale efforts that aimed to reform mathematics teaching as well as transform the broader understanding of the nature of mathematical research. Both reflected the spirit of generalization and abstraction that has been a prominent feature of mathematics since the mid-nineteenth century. By drawing on set-theoretically formulated abstract methods, they helped shape as well as spread new forms of mathematical knowledge that led to important innovations and sub-disciplines. The Soviet school and the Bourbaki movement provided two frameworks for what mathematics ought to be, both of which proved highly influential for the pursuit of mathematical research at several leading centers around the world. Both embodied modernist impulses that developed into major cultural movements shaped by some of the century's most charismatic personalities, talented mathematicians with personal achievements of the first magnitude, but who also expressed clear guiding visions for the future course of the discipline at large. From a philosophical standpoint, both represented viewpoints orthogonal to the foundational issues that engaged many mathematicians in the West, and which led to the three schools of thought known as logicism, formalism, and intuitionism (see Chapter 4). Apart from these shared characteristics, however, the two programs differed quite sharply regarding major issues, such as the essence of mathematical knowledge or its role in science and society. As such, each reflected a distinct vision of what inventing mathematics ought to mean.

Soviet mathematics presents the fascinating picture of complex interaction between an older tradition of excellence extending backward into the Tsarist period and an ideologically committed government that was primarily interested in developing an economy structured in a particular way (socialism/communism). The Soviet Union drew on impressive human intellectual resources both to fulfill its primary social and economic objectives and to enhance its prestige among nations in a flourishing world of mathematics full of opportunities for solving challenging old problems and creating new theories.

Two main centers dominated mathematics in the Soviet Union: Moscow and Saint Petersburg (Leningrad, 1924–91). Each had formidable leaders whose work and methodological preferences strongly colored the research activity in their respective localities. In the wake of the Russian Revolution, both centers came under political pressure to reorganize their educational programs in accordance with the principles of a communist society. They responded, in turn, by adopting eclectic, vaguely Marxist orientations predicated on materialist principles. Above all, Soviet mathematicians excelled in the art of problem solving in work that ranged over problems at all levels of difficulty and that covered virtually every area of mathematical knowledge. They cultivated this special talent by drawing on a wide variety of tools and methods taken from several very different disciplines, including probability theory, number theory and descriptive set theory.

Problem solving, in the concrete sense practiced by Soviet mathematics, played only a secondary role in the Bourbaki project, which undertook an ambitious effort to present an overview of what mathematical knowledge had come to be. With this goal in mind, the group placed a heavy accent on the systematic development of a few fundamental theories. Their members designed these to serve as the basic architectonic foundation for all mathematical knowledge, out of which all future developments should arise. Two main ideas provided the central organizing principles for Bourbaki's image of mathematics: formal axiomatics and the notion of mathematical structures. These general principles guided the work of Bourbaki as the group gradually developed and promoted a universal program that would play an instrumental role in shaping the course of mathematical research.

RISE OF THE SOVIET SCHOOL, 1917–45

Throughout the 73-year existence of the Soviet Union with its varying member states, but also during the decade that followed its disintegration, its mathematicians produced a prodigious amount of first-rate work. Not only did they solve several longstanding problems, they also created new areas of research while greatly increasing the number and quality of mathematical journals and books in the world. While it is difficult to give a complete picture of this "mountain range" of intellectual work, we can get some idea of its extent by looking at the territory on which the "mountains" stand and examining some of the higher peaks.

During the nineteenth century, the center of gravity in the Russian mathematical school was located in Saint Petersburg, where it was associated with the names Pafnuti L. Chebyshev, Andrei A. Markov, and Aleksandr M. Lyapunov. These figures were among the world's leaders in several areas, notably probability and statistics, number theory, approximation theory, and applied mathematics. These areas of mathematics were cultivated to a somewhat lesser extent in the West, and thus the Petersburg school, which took little interest in foundational issues, developed an international reputation as a center for solving concrete problems. This special tradition stood in sharp contrast with the one that later emerged in Moscow during the first decade of the twentieth century under the leadership of Dmitri F. Egorov, a prominent geometer who had absorbed much of what the French and German analysts and set theorists were doing.

The Moscow mathematicians had always been more philosophically inclined than their counterparts in Saint Petersburg—St. Petersburg had a *physico*-mathematical society and Moscow a *philosophico*-mathematical society. The latter group quickly became involved in foundational questions associated with contemporary philosophical discussions. Egorov gathered about him a number of talented students and assistants, and together they investigated delicate questions in measure theory and descriptive set theory. His best student, Nikolai Nikolaevich Luzin, went to Germany, where he studied under Edmund Landau and others. Returning to Moscow, he took his doctoral degree with a brilliant dissertation on integration and trigonometric series.

Luzin immediately gathered about him a group of talented young people, beginning with Pavel S. Aleksandrov and Mikhail Ya. Suslin, who pushed the frontiers of Georg Cantor's continuum hypothesis (abbreviated CH within axiomatic set theory; see Chapter 3 in Volume 5). This hypothesis states that with respect to size (cardinality) every subset of the real numbers falls into one of only three classes: (1) finite sets, (2) countable infinite sets, and (3) uncountable infinite sets with the same cardinality as the entire set of real numbers (the continuum). The crux of the matter for proving CH was case (3), which except for relatively simple subsets of the real numbers remained an open problem. Then, in 1916, Aleksandrov proved it for essentially all the sets required for integration theory (the *Borel* sets). Immediately afterward, Suslin succeeded in extending this result to the larger class of all *analytic sets*. Luzin himself worked in this area for over a decade, finally producing a definitive monograph on the subject, published in French in 1930.[1]

Luzin's school had barely begun to form when the October Revolution, followed by a civil war, disrupted scholarly life in all areas, dispersing scholars for a brief time away from the large cities, where economic life was uncertain. When stability returned in the early 1920s, the former Russian Empire had become the USSR, adding new member republics up through the year 1940. Under the Soviet regime, these new republics acquired their own academies of science and received heavy investment in education through the founding of universities and research institutes. Some places, such as Kharkov, already had a scholarly tradition, so that the Soviet government was not always obliged to start with a blank slate.

Under the leadership of outstanding scholars, Soviet mathematics placed a distinctive stamp on the subject by creating new fields of research and developing new methods for older ones. Among the new fields was point-set or general topology, created largely by Aleksandrov and his collaborator Pavel S. Uryson, who were among the first to study and classify general topological spaces. This theory would later play a major role in Bourbaki's structural approach to modern mathematics. General topology had roots in work undertaken in the late nineteenth century in France and Germany, where mathematicians introduced concepts such as connectedness and continuity independent of a distance measure. Abstract topological spaces thus represented a great generalization of classical geometric spaces based on a Euclidean metric or the manifolds studied by Riemann and others. Subsequently, applications of algebraic methods to explore the properties of these spaces—and prominently among these, homology groups—would become, as shown below, increasingly common.

Point-set topology emerged as a major new field of research in Poland after the Great War. Polish mathematics began to flourish after Poland regained its independence in 1919 and successfully fought off an invasion by the Red Army the following year. Led by Wacław Sierpiński, while drawing on the work of the German topologist Felix Hausdorff, Polish mathematicians achieved several

FIGURE 3.1. East meets West: Participants at the Moscow Topology Conference, September 1935; Pavel Aleksandrov is seated to the far right in the front row next to Heinz Hopf. ETH-Bibliothek Zürich, from Wikimedia Commons as part of a cooperation with Wikimedia CH.

fundamental results in general topology. Sierpiński had participated in the Luzin Seminar in Moscow during the First World War before settling in Warsaw. In the 1930s, as political tensions between the USSR and West became exacerbated, Luzin found it prudent to denounce Sierpiński and cut off all contacts with the Poles.

In the meantime, Luzin's school, whose members facetiously called it "Luzitaniya," published some remarkable work. This group included young talents, such as Aleksandr Ya. Khinchin and Andrei N. Kolmogorov. Luzin, however, had difficulty reconciling himself with the fact that these students were powerful intellects who no longer needed his mentoring. Strains developed and by the late 1920s Luzitaniya had nearly disappeared; only a few loyalists remained faithful to Luzin and continued to work with him. This generational conflict arose just when Stalin attained supreme power over the Soviet government, after which he began to place all scientific activity under his control. Along with the Soviet economy, he wanted to collectivize the entire scientific research/educational system in the USSR. At his instigation, the

Academy of Sciences moved from Leningrad to Moscow. In addition, Stalin wanted to encourage a sense of Soviet identity among scholars by motivating them to publish their works in Soviet languages and in Soviet journals while discouraging scholarly contacts with the West. This "Sovietization" of the academic world gave Luzin's disgruntled students an opportunity to air their grievances against their mentor, even though Stalin himself had no interest in such internecine academic quarrels. Urged on by Party zealots, he nevertheless gave his approval for a public campaign that focused on Luzin as a representative of decadent Western bourgeois science.

Several sets of interests were at stake in July 1936, when the Academy of Sciences held hearings into the accusations made against Luzin.[2] At the center was Luzin himself, still hoping to advance the reputation of his school, even though he had given up teaching at the University. At the hearings, he received support from a few, mostly older scholars, such as Sergei N. Bernstein, head of probability and statistics in the Mathematical Institute of the Academy of Sciences in Leningrad who had entered the "honors class" by solving Hilbert's nineteenth problem (Yandell 2002: 374–7). Arrayed against Luzin was, first, a large and vocal group of his former students. They clamored for his removal from the Academy of Sciences, convinced that he had held them back in their careers while stealing their work. The most strident of these several voices was that of Aleksandrov. Supporting the students were the Party ideologues, led by the Marxist philosopher/mathematician Arnošt Kolman and Lev Z. Mekhlis, the editor of *Pravda*. Both promoted a purely Marxist approach to research and education.

Finally, behind the scenes and directing the action was the Politburo, represented by Stalin and Molotov, who were trying to reorganize research and education into a system run from the top down by the higher Party officials. They communicated their wishes to Gleb M. Krzhizhanovskii, an old Bolshevik engineer, who had been in charge of the electrification project of the 1920s and early 1930s. He now assumed the job of chairing all but the first of the hearings at the Academy. These lasted up until mid-July and thus ended just before the first major political purges of the late 1930s. Luzin narrowly escaped prison or worse, but in the end kept his position in the Academy of Sciences. This rupture with all but a small remnant of his former students, however, was irrevocable from that time until his death.

Much of what makes Soviet mathematics distinctive came about during the 15-year-period from the Great Breakthrough (*Velikii Perelom*) of 1929, marked by the first five-year plan, to the end of the Second World War in 1945. Not only mathematics, but all aspects of Soviet life were "Sovietized." During this period, the unique character of Soviet mathematics emerged with works that made a long-lasting mark on a wide variety of research areas. These developments revealed, indeed, that the talents of Luzin's students went far beyond descriptive set theory and point-set topology. Three specific examples illustrate how Soviet mathematicians deepened the foundations in one established field, contributed to the solution of a famous problem in another, and proved a seminal theorem that effectively launched an entire new field of research.

In the late 1920s, Khinchin and Andrei N. Kolmogorov published groundbreaking papers on modern probability theory, work that culminated in 1933 when Kolmogorov laid down axioms for a probability space. That foundational result solved part of the axiomatization program Hilbert had set forth in his sixth Paris problem (Corry 2004: 164–7). A second major achievement in this period was a partial solution of the famous Goldbach Conjecture, which asserts that every even number is the sum of two prime numbers. In 1937, Ivan M. Vinogradov showed that every sufficiently large odd number can be expressed as the sum of three prime numbers. Vinogradov published this epoch-making paper, written in Russian, in a Soviet journal, an indication that Stalin's campaign to Sovietize scholarship in the USSR had real effect.

A third Soviet contribution to higher mathematics had a profound impact on a research field still in its infancy. In 1941, Izrail' M. Gel'fand published a paper, written in German, in the Soviet journal *Matematicheskii Sbornik* that marks a major breakthrough in functional analysis. This subject had arisen in the first decade of the twentieth century in the wake of Hilbert's work on integral equations. Through publications of his Göttingen school the idea gained currency of regarding a function as a point in a vector space (Hilbert space). There were two major classes of such spaces, one based on continuous functions, the other on integrable functions. If such a space happens to have a natural multiplication operation in addition to its vector space structure, it is then a *Banach algebra*. Gel'fand called it a *normed ring*, and he showed that its algebraic structure could be used to create a topological space (called the *maximal ideal space* of the algebra). He showed, moreover, that one could represent the original algebra as an algebra of continuous functions on that topological space.[3]

Gel'fand's seminal paper made essential use of the point-set topology that Polish and Russian mathematicians had developed over the previous two decades. In particular, the method of defining a topological space using a closure operation, invented by the Polish mathematician Kazimierz Kuratowski, turned out to be perfectly adapted to the algebraic structure of a Banach algebra. Thus, one abstraction (a topological space) played a vital role in linking a second abstraction (a Banach algebra) to classical theories of great interest, such as the theories of Fourier and Laplace transforms. This new approach generalized these familiar types of transforms by making use of concepts from modern algebra (ideals and homomorphisms) and point-set topology. The consequences were profound, providing new proofs of some classical theorems such as

FIGURE 3.2. Andrey Nikolaevich Kolmogorov at the Institute Henri Poincaré, September 1958. Photo by Keystone-France / Gamma-Keystone / Getty Images.

Norbert Wiener's Tauberian Theorem. Banach algebras offered a unified point of view in other established areas along with many completely new results in topology and analysis. It soon emerged as a major topic of research in the area of functional analysis. Centers specializing in this field arose in places as far separated as Yale University in the United States—led by Charles Earle Rickart—and Tokyo, Japan, where Mito Nagumo had defined the concept of a Banach algebra as early as 1936, using the nomenclature *linear metric space*.

These three examples illustrate the worldwide influence of Soviet mathematics, but, in fact, one could name many others in nearly all areas of mathematics. Moreover, in mathematical logic and foundations of mathematics,

Soviet scholars defended a distinctive position very far from the practical, applied mathematics that Marxism might seem to demand. Nor did they completely ignore the debates among logicists, formalists, and intuitionists that for a time dominated foundations discussions in the West. For example, during the late 1920s, Khinchin and Valerii I. Glivenko took part in debates over intuitionism, published in the *Bulletin of the Belgian Royal Academy of Sciences*. They argued against the view that—since a statement could be true, false, or undecidable—intuitionism was a propositional calculus based on three-valued logic. Khinchin showed that one proposal to rescue intuitionism from this three-valuedness was not feasible, whereas Glivenko showed that intuitionism was, in fact, not three-valued, ultimately proving that the statement "*p* is neither true nor false" is false.

The German invasion of 1941 and the ensuing war fought mainly on Soviet territory, led to privations and disruption far exceeding that of the Civil War that followed the Revolution. Yet despite the foreign occupation of a considerable portion of the country, the Academy of Sciences, meeting in Sverdlovsk in 1942, confidently formulated plans for rebuilding these areas after the war. All the major academies and universities in Russia, the Baltic States, the Caucasus, and Central Asia relocated as necessary, bringing with them what they could of their physical plant. For example, a significant portion of Moscow University moved to Ashkhabad (Ashgabat) in Turkmenistan. Thus, a semblance of normal academic life went on, even as the war raged along a front 2000 kilometers long. The Ukrainian Academy of Sciences relocated to Ufa, and there Nikolai N. Bogolyubov made fundamental advances in nonlinear and statistical mechanics. He was later to play a major role in the development of nuclear weaponry in the USSR. The involvement of many Soviet mathematicians in military research proved to be an important factor during the post-war period.

MODERN ALGEBRA AND THE RISE OF BOURBAKI

Whereas Soviet mathematics had eclectic roots and drew heavily on new fields of research spawned by Cantorian set theory and modern analysis, the French mathematicians who founded the Bourbaki movement took much of their inspiration from parallel developments in the field of algebra. Throughout the nineteenth century, algebraic research had a quite concrete character. Algebraists were concerned with the theory of polynomial equations and polynomial forms, including algebraic invariants. They introduced and used modern concepts, such as groups and fields, but only rarely investigated these as topics of inherent interest. Moreover, they saw these notions as tools rather than as fundamental foundational concepts that might be used to reconceptualize the larger edifice of algebra. Groups were studied as groups of substitutions in the context of equation theory or what came to be called Galois theory (see

Chapter 3 of Volume 5). Fields were studied mainly as extension fields of the rational numbers $\mathbb{Q}$ or as subfields of the complex numbers $\mathbb{C}$.

This classical disciplinary image emerges clearly in Heinrich Weber's three-volume *Lehrbuch der Algebra*, published toward the end of the century. In this emblematic work, abstract concepts such as groups were subordinated to the main tasks of classical algebra, namely to answer questions related to polynomial equations and forms. Weber's text stands in sharp contrast with the textbook that replaced it thirty years later, Bartel L. van der Waerden's *Moderne Algebra*, a two-volume survey published in 1930–1. This work provided a comprehensive synthesis of abstract algebraic structures using modern axiomatic methods. Its author had learned these ideas in Germany, where he took courses taught by Emmy Noether in Göttingen and by Emil Artin in Hamburg.

This largely German approach to algebra eventually took on central importance for the Bourbaki movement, though the original group had formed with a different motivation in mind, namely, to modernize the French mathematical curriculum, in particular in analysis. This project, however, gradually gave way to a much broader one that constituted a new vision for mathematical knowledge as a whole. This entailed a new conception not only of algebra but of several other disciplines as well. Accordingly, the Bourbaki group promoted a clearly defined agenda for what mathematical research should look like in the future with ideas that had a momentous influence on hordes of mathematicians around the world. The principal vehicle they used to communicate this vision was a multi-volume series published between 1939 and 1998 under the title of *Eléments de mathématique*.

Among the founding members of the Bourbaki group, the most influential and active were Henri Cartan, Claude Chevalley, Jean Dieudonné, and André Weil—all former students of the École normale supérieure in the early 1920s. Over the years, many younger, but distinguished mathematicians participated in the group's activities, while the elder members were (theoretically) supposed to drop out by the age of fifty. Among younger-generation Bourbaki members, some of the most prominent included Samuel Eilenberg, Laurent Schwartz, Armand Borel, Jean-Pierre Serre, Serge Lang, Alexander Grothendieck, Pierre Samuel, Pierre Cartier, and Jean-Louis Verdier. These figures were some of the leading mathematicians of their generation, actively pursuing their own individual research in different specialties, while the activities of Bourbaki absorbed a good part of their time and energy.

In the early stages of their academic careers, these young men felt dissatisfied with the traditional French course in analysis, as reflected in textbooks written by older masters: Jacques Hadamard, Emile Picard, and Edouard Goursat. In addition, they felt that postwar French mathematics was lagging behind that of other countries, especially Germany. This situation provided the central motivation for the deliberations that would lead to the Bourbaki project, namely, to search for better ways to introduce young students into the basic concepts and theorems in modern analysis. They started by meeting in Paris, arriving on a weekly basis from the provincial universities where they began their careers. Their meetings took place at the *Séminaire de mathématiques*, founded in 1933 at the Institut Henri Poincaré by Gaston Julia, then a senior figure in the French mathematical community. Each academic year, this seminar was devoted to a single general area of research. By focusing on one area at a time, participants hoped to gain a broader and more systematic knowledge of recent developments in the various branches of pure mathematics. For each meeting, one of the participants was responsible for preparing the topic under discussion, afterward editing his talk and then distributing it to the other participants.

By December 1934, an ambitious idea started to take shape among the group, namely, to produce a treatise that would define the syllabus for the certificate in differential and integral calculus for decades to come. Of central importance was the explicitly stated aim that this treatise should be "as modern as possible." Once the project started to materialize, however, it soon became apparent that this aim, however suggestive, stood in need of further articulation. The participants soon realized that recent developments in disciplines that departed from the knowledge classically associated with the calculus in the French tradition should become a necessary component in the project. These included modern algebra as presented in van der Waerden's book; the theory of integral equations, with special emphasis on the Hilbert space; and recent developments in the theory of partial differential equations. Sub-committees were then formed to oversee the writing of the various parts of the treatise, and the participants agreed that final decisions on its contents had to be reached by general consensus.

Another important aspect concerned the expository principles adopted by the group from very early on. Perhaps the most significant of these pertained to general results needed to prove specific theorems: Bourbaki insisted on developing these independently rather than in the course of the exposition itself. This approach ran counter to the presentations in classical French texts on analysis, which took up basic ideas one after the other in extremely detailed fashion, but rarely considered general concepts that could account for many different cases. These young French mathematicians not only felt dissatisfied by the standard of rigor in these older presentations of the calculus, they also thought that general, abstract theories needed to be developed first and then applied later, wherever needed. Over time, this principle led to elaborate discussions of general abstract concepts, results and proofs. In many cases, the motivation behind this theorizing was far from transparent to prospective readers of the treatise.

The first real working meeting of the Bourbaki group took place in July 1935 in the village of Besse-en-Chandesse, close to Clermont-Ferrand. According to

FIGURE 3.3. The Bourbaki Congress in Dieulefit 1938, (*left to right*) Simone Weil, Charles Pisot, André Weil, Jean Dieudonné (sitting), Claude Chaubaty, Charles Ehresmann, and Jean Delsarte.Courtesy of the Association des collaborateurs de Nicolas Bourbaki.

the original plans, they hoped to complete the entire treatise within one year. No one imagined this as a project that might go on for decades. Along with classical themes in analysis, this initial group gave increased attention to basic notions in algebra, topology and the theory of sets. These now appeared necessary to provide the presentation with the kind of coherence and modern perspective that the group insistently spoke about. Up until the outbreak of the war, they continued to organize "congresses" three times a year in different places around France for one or two weeks. At each meeting, individual members distributed drafts of the various chapters, which the other members usually criticized harshly and then reassigned for revision. Only after several drafts was the final document ready for publication. Each chapter and volume of Bourbaki's treatise was thus the outcome of arduous collective work, making the spirit and viewpoint of the person(s) who originally wrote it hardly recognizable. Despite rather scanty documentary evidence, the personal dynamics within the Bourbaki group are a matter of considerable interest, as it represents, no doubt, a nearly unique case in the history of science. In spite of appearances, the activities of the group throughout the years and the results of their joint work were anything but a monolithic undertaking. Indeed, the ideas behind the entire project underwent complex processes of elaboration and evolution. The most surprising fact about Bourbaki, so it seems in retrospect, is that it could work at all.

As the treatise gradually took on form, it eventually outgrew the original framework for a modern textbook in analysis. Instead, it evolved into a multi-volume treatise, entitled *Eléments de Mathématique*, each volume containing a comprehensive exposition of a different mathematical discipline. Most of the material covered was not in itself new and, very much like with van der Waerden's *Moderne Algebra*, the novelty came about through the organization

FIGURE 3.4. An early meeting of the Bourbaki group. Standing (*left to right*): Henri Cartan, René de Possel, Jean Dieudonné, André Weil, and Luc Olivier (a biologist); seated: Mirlès, Claude Chevalley, Szolem Mandelbrojt. Courtesy of the Association des collaborateurs de Nicolas Bourbaki.

of the mathematical knowledge, which stressed underlying unity and thereby conveyed a new overall conception of mathematics. Initially, Bourbaki thought to begin the opening chapters with a *paquet abstrait*, initially conceived as a supporting toolbox of limited scope, but these tools gradually assumed central stage and became the central core of the treatise. This caused the group to delay introducing classical topics in analysis, some of which were eventually relegated to specific sections, while others were left as exercises or even left out of the treatise altogether.

The first chapter of the *Eléments* appeared in 1939. By this time, the plan had settled around six basic books: I. Theory of Sets; II. Algebra; III. General Topology; IV. Functions of a Real Variable; V. Topological Vector Spaces; VI. Integration. At a second stage in the 1950s, additional chapters were added to this plan; these included, Lie Groups and Lie Algebras; Commutative Algebra; Spectral Theories; Differential and Analytic Manifolds (which was essentially no more than a summary of results). In its final form, the treatise comprised over seven thousand pages, with new chapters continuing to appear until the early 1980s.

When the group resumed activities after the hiatus imposed by the war, some changes were noticeable and they were manifest at two separate levels. On the one hand, there was the treatise, which continued to be a principal focus of attention and to embody the vision of what the basis of mathematics was about, and what any person intent on moving into the active areas of advanced research should first learn. On the other hand, there was a new seminar started in Paris by Cartan in 1948–9, which was in a sense the revived version of the old *Séminaire Julia*. The *Séminaire* became a coveted stage for presenting and discussing innovative work, mostly done along the line of individual interests of current Bourbaki members. With time, this highly dynamic seminar (which would eventually evolve into the famous *Séminaire Bourbaki*) turned out to be a much more adequate place for the continued discussion of groundbreaking ideas than a treatise intended as a foundational text that should continue to be relevant for decades to come.

As the collective work of the Bourbaki group continued to thrive in the post-war era and the treatise took further shape, an implicit but pervasive notion increasingly came to the fore as the guiding principle underlying their work. This was their conception of mathematics as a systematic, elaborate hierarchy of structures, which served as a framework not only for understanding present-day mathematics but also for guiding the invention of future mathematical knowledge. This vision was essentially an extension to the entire realm of pure mathematics of the idea that underlies van der Waerden's textbook for the more circumscribed case of algebra. Bourbaki's treatise presented not only algebra but also topology and functional analysis, as well as the other topics, as individual manifestations of the same general idea: the notion of a mathematical structure. This conception implied that research in all these fields should be undertaken from a unified methodological perspective, mediated by ideas such as isomorphisms, quotients, extensions and the like. However, even though it dominated the entire Bourbakian enterprise, this notion of a mathematical structure was *not* a formally defined concept within an elaborate mathematical theory. Its purpose was rather to provide a substantial answer to an increasingly pressing question. Faced with the unprecedented growth and diversification of the discipline, Bourbaki's vision offered an answer to the question of the "unity of mathematics."

In 1948, under the name of Bourbaki, Jean Dieudonné wrote an article that explicitly discussed these fundamental ideas. Published in English in 1950, "The Architecture of Mathematics" was later called the group's manifesto, which Dieudonné described in these words:

> At the center of our universe are found the great types of structures, ... they might be called the mother structures . . . Beyond this first nucleus, appear the structures which might be called multiple structures. They involve two or more of the great mother-structures not in simple juxtaposition (which would not produce anything new) but combined organically by one or more axioms, which set up a connection between them . . . Farther along we come finally to the theories properly called particular. In these the elements of the sets under consideration, which in the general structures have remained entirely indeterminate, obtain a more definitely characterized individuality.[4]

Alongside the notion of structure, the view that mathematics was indeed a strongly unified branch of knowledge in spite of appearances relied on the sweeping adoption of the modern axiomatic method as the natural framework for the presentation of mathematical knowledge. In doing so, Dieudonné frequently described Bourbaki as the 'natural heir' and follower of David Hilbert, who opened this path at the beginning of the twentieth century with his seminal work on the foundations of geometry and arithmetic. In fact, there were very significant differences between Bourbaki's and Hilbert's respective conceptions of the nature, meaning and scope of the axiomatic method, though from Bourbaki's perspective the austere, non-motivated style of presentation was (wrongly) seen as the essence of Hilbert's putative program for axiomatizing all of mathematics. The *Eléments* thus put forward a unified, comprehensive picture of what the Bourbaki group saw as the main core of mathematics. They adopted a standard system of notation, asked similar questions about the various fields investigated, and introduced similar tools and methods across apparently disparate mathematical domains. Hence, this "structural" approach proved very successful as an *informal* image of mathematics at large.[5]

SOVIET VS. WESTERN MATHEMATICS

Throughout the twentieth century, the predominant view of the vast majority of Western mathematicians regarding foundational structures assumed the validity of Cantorian set theory (as in Book I of Bourbaki's *Elements of Mathematics*). They therefore accepted that the principal structures of modern theories were infinite sets, pictured as collections of simultaneously existing objects. In contrast to finite collections, such sets have no prototypes in the real world, leading some to doubt their legitimacy. These doubters have typically advocated a constructivist approach to mathematics, one example being recursive function theory, an approach that raises the possibility of constructing mathematics without resorting to the concept of an actual infinity.

The Bourbaki program, which set forth an all-embracing framework for the entirety of mathematics, necessarily represents a very high level of abstraction. Yet, going back to the nineteenth century, when abstraction was rising on all fronts, there have always been countercurrents. The best-known example in the twentieth century arose in the 1920s with intuitionism, whose most famous champion was Luitzen E. J. Brouwer. Another countercurrent to the mainstream approach to foundations of mathematics is constructivism, which is less elegant than the standard presentation of mathematics but has the advantage that the objects it works with have greater "palpability."

The founder of the constructivist school in the USSR was the son of the famous Andrei A. Markov, also named Andrei A. Markov; he received his doctoral degree in 1921, the year before his father died. Markov became interested in foundational questions after the Second World War. In 1958, he began writing about constructivism and continued to do so up until his death. Those who followed his lead and took a constructivist approach were pragmatically oriented; they had no use for the ethereal concepts and methods of Western foundational programs promoted by logicists and formalists. Their core field of interest was analysis with strong emphasis on approximation theory. One of the major figures in this area was Kolmogorov's student, the very long-lived Sergei M. Nikol'skii. From a general foundational standpoint, the Soviet school sought to reformulate mathematical analysis using recursive methods in order to avoid the more controversial aspects of set theory, especially actual infinities, the law of excluded middle, and the axiom of choice. During the early decades of the century, all three of these notions became standard tools in the arsenal of mathematicians in the West, who were loath to give up such fundamental results as, for example, the intermediate value theorem.

Intermediate value theorem This says that a continuous real-valued function f defined on a closed interval [a.b] with f(a) = c and f(b) = d will assume every value in the interval [c,d]. Not only is this claim intuitively obvious, the proof is so simple that one can find it in nearly every elementary calculus textbook. The trouble with this theorem, though, is that the claim it makes is entirely nonconstructive. Thus, if one takes a number t in the interval [c,d], then we can prove there must exist a number s in [a,b] such that f(s) = t because if this were not the case, then f could not be a continuous function. Were we to name a specific function f along with an interval [a,b] and a number t in the interval [f(a), f(b)], the intermediate value theorem would have nothing at all to say about how we could find, or even approximate a number s in [a,b] with the property that f(s) = t.

Zermelo's Axiom of Choice (AC) asserts that from a collection of subsets of a set X, one can freely choose one element from each of these subsets. This "choosing" takes place "all at once," that is independent of time. In 1919, the philosopher Bertrand Russell gave a vivid example to explain when a mathematician needs the axiom of choice, updated here for a twentieth-first-century reader. One must first imagine an incredibly wealthy person *M*, who happened to own infinitely many ordered pairs of shoes and socks. *M* demands that a servant ε pick out one shoe and one sock from these two infinite sets. As it happened, ε had studied set theory, so she explained that this task presented a subtle mathematical difficulty. Picking out one shoe from each pair would be easy, since she could employ an algorithm that determined whether to choose the right or left shoe. There was no such method available, however, for choosing between the socks, so for these it would be necessary to employ the axiom of choice.

In the constructivist approach, on the other hand, one regards a real number as a computable sequence of decimal digits. Thus, a constructive number is a "constructive object"—a notation for a program (algorithm) that computes the corresponding recursive function or, still better, the index of that notation in a specific enumeration. Similarly, a constructive function is based on the indices of certain computable functions. This area received considerable development among Soviet mathematicians during the 1960s, when certain basic principles of constructivism were developed. Thus, constructive analysis differs from standard analysis in fundamental ways, and some Soviet mathematicians worked on finding constructive analogues for the standard objects of analysis. They came up with constructive definitions for a host of fundamental concepts in the theory of measure and integration. They also sought constructive versions of classical notions like the Riemann integral, analytic functions of a complex variable, topological spaces, and the Brouwer fixed-point theorem. Work in this area continued throughout the life span of the Soviet Union.

Both constructivists and intuitionists had long objected to the fact that so much of modern mathematics—including all that relies on the axiom of choice—takes place "offstage," with no hint of how to implement the results. Hermann Weyl spoke of mathematical theorems issued in the form of worthless banknotes during a period of hyperinflation. The many pure existence proofs, for example, became a kind of "black box" whose contents remained hidden. Mathematicians who relied on the law of excluded middle to prove existence theorems compounded the offense against intuitionism and constructivism. Brouwer harshly attacked Hilbert's unrestricted use of this principle, referring to it as "Hilbert's dogma" (see the Introduction).

Intuitionism and constructivism seemed to have much in common, since both rejected the *deus ex machina* provided by the axiom of choice and the law of excluded middle. Seen in Marxist terms, however, intuitionism was nevertheless idealistic and tended to hypostatize mathematical concepts in a rather Platonist manner. The Marxist/constructivist approach was grounded in computations that, in principle, could be implemented on a powerful computer, more powerful than is likely to be available at a given moment. Soviet mathematicians, as noted above, had participated in debates over the proper foundations of mathematics in the 1920s, and they found ample grounds for keeping intuitionism off center-stage. As early as 1925, Kolmogorov explored the limitations of the law of excluded middle. To each logical formula, he assigned the formula obtained by replacing every sub-formula in it by its double negation. He then showed that when the original formula is deducible in classical logic, then the transformed formula can be deduced using intuitionistic logic and vice-versa. In this way, he obtained an intuitionistic interpretation of classical logic.

Another response of some Soviet mathematical philosophers and logicians involved the application of dialectical materialism. A leading champion of this "dialectical" approach was Kolman, who wrote many books and articles claiming that "bourgeois" mathematics was in a state of crisis. He saw this as a classical example of the Marxian "thesis/antithesis" dialectic, in this case an instance of opposition between continuous concepts (geometrical lines, real numbers, and continuous mappings of them) and discrete concepts (integers, finite groups, and discontinuous functions); both had arisen to occupy an equally prominent place in the world of mathematics. He insisted that the "bourgeois" mathematicians had gotten themselves into an impossible position, from which the only escape would be to synthesize these two following Marxist dialectical principles. While Kolman's 1936 book on mathematical methods received very poor reviews from the Soviet mathematical community, the "dialectical" party in Soviet mathematics continued to promote its ideas right up to the end of the Soviet Union. The Marxist approach to mathematics

FIGURE 3.5. Lomonosov Moscow State University. Photo by Jarana Creatives / Getty Images.

naturally and legitimately engaged the attention of historians of mathematics, such as Konstantin A. Rybnikov, though it never became a major area of research for the vast majority of Soviet mathematicians.

After the war, the leaders of the USSR were primarily concerned with repairing the damage it had caused while developing nuclear weapons to break the American monopoly. The latter effort proved successful, and by 1960 the country's space program appeared to have surpassed that of the United States. As for Soviet mathematics, it had more than earned the respect of Western mathematicians. In 1951, the American Mathematical Society began a series of translations of Soviet mathematical papers, followed by a second series in 1955, which eventually provided cover-to-cover translations of the major Soviet journals. This interest was reciprocated by the other side, as in 1952, the Soviet Academy of Sciences began publishing review journals in a number of areas of science. All of these appeared under the general rubric of "reference journal" (*referativnyi zhurnal*), one of the areas being mathematics. Thus, from that date on, Soviet mathematicians enjoyed the use of a reference tool comparable to *Zentralblatt für Mathematik* and *Mathematical Reviews*.

While Soviet mathematicians played important roles in the nuclear and space programs, they also continued to make fundamental advances in nearly every area of mathematics, including mathematical physics, analysis, ordinary and partial differential equations, algebraic topology, mathematical logic, and countless others. Here we mention just two examples from particular areas of research. In classical complex analysis, a standard result states that an analytic function has a local representation as an infinite Taylor series, which converges for all points inside a disk of known radius. In 1951, Sergei N. Mergelyan proved that one can replace these infinite series with (finite) polynomials by showing that any analytic function can be approximated by polynomials defined on sets which are geometrically more complicated than disks. This remarkable theorem is analogous to the Weierstrass approximation theorem for continuous real-valued functions on a finite closed interval of the real line. Another major advance occurred in the area of representations of abstract groups and harmonic analysis (Fourier transforms) based on the earlier work of Gel'fand, discussed above. In 1934, Lev S. Pontryagin established a fundamental duality theorem between any abstract group and its so-called *character group* by showing that the character group of the character group was structurally identical to the original group.[6]

The communist emphasis on economic development gave a distinctly applied and practical bent to a great deal of Soviet mathematics. Perhaps the most "theoretical" area developed extensively in the USSR, beginning toward the end of the Second World War, was theory of algorithms, a subfield of algebra. Algorithms serve as the main nexus between abstract algebra and the concrete world of computer applications. Other areas of mathematics with an applied flavor were optimal control theory and cybernetics, both of which took root in the countries of the former USSR, producing significant results right up to the present day.

The Soviet Union itself reflected the philosophical views of Marx, Engels, Lenin, Stalin, and others. It remains difficult, however, to assess the extent to which the Marxist state ideology influenced Soviet mathematicians. A popular three-volume set of articles argued that mathematics actually offers a proof of dialectical materialism. Aleksandr D. Aleksandrov wrote this telling passage:

> Logicism, intuitionism, formalism and other such systems emphasize one or another piece of mathematics – its connection with logic, intuitive clarity, formal rigor, and the like – and without proper grounds exaggerate and absolutize its significance, detach it from reality, and after an intensive analysis of that one feature of mathematics in isolation, lose sight of mathematics as a whole. It is because of this one-sidedness that none of these systems, despite the subtlety and profundity of some particular deductions, can lead to a true understanding of mathematics. In contrast to the various trends and shades of idealism and metaphysics, dialectical materialism examines mathematics and all sciences as they are, with all the richness and complexity of their connections and evolution. And this is because dialectical materialism strives to understand all the richness and complexity of the

> connections between science and reality, the whole complexity of its growth from simple generalization of observation to the highest abstractions and thence to practical application, because it constantly aligns its very approach to science with the objective content and new discoveries of science. In the final analysis, it is for that reason alone that this is the only objectively scientific philosophy that leads to a true understanding of science in general and mathematics in particular.[7]

The editors also argued that the axioms of arithmetic and geometry clearly had to be consistent, since they corresponded to real structures that satisfy them. That is, these mathematical structures can be used in descriptive models of the physical world whose consistency with observation can be determined empirically. Such a claim stood in sharp opposition not only to the position of Bourbaki but also to mainstream Western views since the time of Hilbert. Those who espoused modern axiomatics held that mathematics is concerned exclusively with formal theories, which derive valid statements from general principles using specified rules of inference.

FIGURE 3.6. A meeting of the Bourbaki group after the Second World War, (*left to right*) Roger Godement, Jean Dieudonné, André Weil, Saunders Mac Lane, Jean-Pierre Serre. Courtesy of the Association des collaborateurs de Nicolas Bourbaki.

In the decades following the Second World War, Bourbaki's image of mathematics was dominant in France. Several leading members of the group, notably Chevalley, Weil and Dieudonné, also worked abroad at other important centers of activity, mainly in the USA and in Brazil, and thus helped to spread the gospel further. While it is difficult to assess the extent to which the various parts of Bourbaki's *Eléments de mathématique* were used in either introductory or advanced courses, the concepts and main problems in the treatise clearly had a major impact on many areas of pure mathematics, in particular those most deeply affected by modern algebra. In those fields, numerous texts adopted Bourbaki's nomenclature and idiosyncratic style. Notations such as the symbol e for the empty set, and terms like injective, surjective, and bijective, and concepts such as filters and ultrafilters, owe their widespread popularity to the *Eléments de mathématique*.

It seems remarkable, however, that for all its impact, the *Eléments* never became a textbook of choice for analysis, as originally intended by the founding members. Nevertheless, a few widely used textbooks that became emblematic in the second half of the twentieth century clearly reflect Bourbaki's image of mathematics. This is certainly the case for Dieudonné's textbooks in real and functional analysis, particularly *Foundations of Modern Analysis*, based on lecture notes from courses taught at Northwestern University in 1957–9. In the preface to this book, Dieudonné expressed the essence of modern mathematical thought as conceived by Bourbaki. He thus wrote:

> The fundamental concepts of function theory and of calculus have been presented within the framework of a theory which is sufficiently general to reveal the scope, the power, and the true nature of these concepts far better than is possible under the usual restrictions of "classical analysis." . . . This has also the necessity of a strict adherence to axiomatic methods, with no appeal whatsoever to "geometric intuition," at least in formal proofs: a necessity we have emphasized by deliberately abstaining from introducing any diagram in the book. My opinion is that the graduate student of today must, as soon as possible, get a thorough training in this abstract and axiomatic way of thinking, if he is ever to understand what is currently going on in mathematical research. This volume aims to help the student to build up this "intuition of the abstract" which is so essential in the mind of a modern mathematician.[8]

Something similar could be said of Serge Lang's textbooks on linear algebra, abstract algebra and the calculus, as well as of those of Emil Artin. Artin had been Lang's advisor in the 1950s and also his influence on van der Waerden in the late 1920s had been crucial for the consolidation of the structural approach exposed in the latter's textbook. Artin was himself a distinguished author of textbooks whose style and Bourbaki's strongly resonated with each other, and the publication of the *Eléments* led him to further enhance many of the basic structural components that had informed his writing from very early on. In reviewing Bourbaki's volume on algebra in 1953 for the *Bulletin of the AMS*, Artin praised the "mercilessly abstract" approach followed in this "monumental work," meant to present a comprehensive picture of present day mathematics. He stressed in particular the fact that:

> this exposition is done in such a way that the common bond between the various branches of mathematics becomes clearly visible, that the framework which supports the whole structure is not apt to become obsolete in a very short time, and that it can easily absorb new ideas. Bourbaki achieves this aim by trying to present each concept in the greatest possible generality and abstraction. The terminology and notations are carefully planned and are being accepted by an increasing number of mathematicians. Upon completion of the work a standard encyclopedia will be at our disposal. (Quoted in Rowe 2018: 402)

In the long run, the branches upon which Bourbaki exerted the deepest influence were algebra, topology and functional analysis. These became the backbone of mathematical curricula and research activity in many places around the world, lending these fields prestigious status within the general economy of mathematical knowledge. On the other hand, these trends were not solely due to Bourbaki's prominence and influence, nor did Bourbaki's style set the tone everywhere. Early developments in the USA, for example, had been strongly oriented since their inception towards axiomatically based research in fields like point-set topology or the theory of algebras and hyper-complex numbers, and somewhat later in algebraic topology. The kind of topological research broadly conducted in the USA in the postwar era was far closer to the school of Robert Lee Moore than to Bourbaki's approach. Likewise, similar trends were recognizable in the work of Soviet mathematicians in set theory and real analysis, and in the Polish school of analysis. Yet, the influence of the Bourbaki members, both because of the scope of their individual achievements and the direct and indirect influence accorded to their treatise in so many places, became a truly major factor affecting these processes.

At the same time, disciplines like logic, probability, and most fields of applied mathematics, which were beyond Bourbaki's scope, became underrepresented in the many places worldwide where Bourbaki's influence was most strongly felt. This was the case for many French and several American universities at various times between 1940 and 1970. Further, group theory and number theory, despite being strong points of some members (notably Weil for number theory) were not independently treated in the *Eléments*. Topics like the classification of finite groups and many important problems in number theory were either less amenable to the kind of systematic, comprehensive treatment presented in the collection or considered to be part of the more advanced topics of mathematics. In addition, as part of an underlying tendency of estrangement from the visual, as expressed by Dieudonné in the above quotation, classical geometry was completely omitted from the Bourbakian picture of mathematics, except for what could be reduced to linear algebra.

In trying to understand both the great success of the project and the difficulties that eventually arose as it continued to develop in the late 1950s it is useful to consider two main axes of internal tension that marked the activities of the group. First, there was the basic tension between the collective work of the group and the individual research interest of its members. This produced a natural conflict between the attempt to present a unified perspective of the entire discipline, on the one hand, and the various idiosyncrasies of the extremely original minds involved in the project, on the other. The second axis reflected intergenerational tensions. The younger generation of mathematicians who joined the group beginning in the 1960s, including such rising stars as Jean-Pierre Serre and Alexander Grothendieck, broke with the style practiced by the founding fathers in some important respects.

A main point of contention concerned the language of functors and other concepts from category theory and the role these notions should play within the treatise as well as in mathematical research overall. Indeed, between the publication of the *Fascicule* in 1939 and the four chapters of *Set Theory* in 1954–7, category theory appeared on stage, through the work of Eilenberg and Saunders Mac Lane. Before long, the language of functors and natural transformations started to be universally adopted in fields like homological algebra and algebraic topology. Discussions concerning the latter discipline, to be sure, occupied the members' attention as early as 1946, when Eilenberg joined the group and prepared together with Weil a very detailed report on it. A section dealing with this discipline, however, was never to appear as part of the treatise, whereas the focus of the enormous interest that it aroused among members of the group moved away from the treatise and the debates surrounding it and became central to the meetings the new *Séminaire*.

In 1956, two Bourbaki members, Cartan and Eilenberg, published their textbook, *Homological Algebra*, which was to become a classic in the field. By this time, categories and functors had become the standard framework for treating questions related to homological concepts. Previously, in 1952, Eilenberg had also published a textbook on *Algebraic Topology*, this time with Norman Steenrod, also fully couched in categorical language. Bourbaki's book on *Homological Algebra*, in contrast, appeared only in 1980 and, for that matter, only as chapter 10 of the volume *Algebra*. Adopting the categorical language in the treatise might have been natural, but this would have involved making major changes in sections of the treatise published long ago. In any case, Bourbaki opted to forego such an undertaking. Moreover, not all members of the group were convinced of the potential gains offered by the categorical point of view. Hence, Bourbaki introduced the basic concepts of homological algebra following the narrower, though by then essentially obsolete framework of modules. Thus, whereas Bourbaki's treatment of general topology in the 1940s had embodied a truly innovative approach that many others were to follow, this was hardly the case with homological algebra in the 1980s.[9]

Alongside these conceptual challenges came generational tensions, embodied particularly in the personal as well as mathematical differences between Weil and Grothendieck, who respected each other highly but could not get along easily together. Weil had been the first to suggest that members should retire from active participation at the age of fifty, but ironically, on arriving at that age in 1956, he gave little sign of wanting to diminish his influence on the project. The younger Grothendieck, a highly individualistic personality even when seen from the perspective of this unique bunch of talents, favored the continuation of the generalizing spirit that permeated the early books, but with more powerful, increasingly abstract algebraic tools. Weil was not enthusiastic about Grothendieck's views, nor were other members, including some of Grothendieck's own generation. In 1958, he left the group and continued his own, highly idiosyncratic way (on his subsequent career, see Chapter 4). An anonymous text, which circulated within the group in the early 1960s, dramatized this situation in the following terms:

> I have learnt that Grothendieck is no longer a member of Bourbaki. I regret that very much, as I regret the circumstances that led to this decision . . . [namely,] a systematic opposition, more or less explicit depending on this or that person, against his mathematical point of view, and especially against the use of the latter by Bourbaki. . . . It is a scandal that Bourbaki not only did not take the lead in the functorial movement, but rather that is not even in its tail. . . . If some of the founding members (e.g., Weil) wish to revert on the decision not to influence the direction that Bourbaki wants to follow, he should say so explicitly. . . . If Bourbaki refuses, not just to join the new movement, but to take the lead in it, then those treatises pursuing the formulation of the elements of mathematics (and not just those dealing with algebraic geometry) will be written by others who will take inspiration not in the spirit of Bourbaki 1960, but in his spirit 1939. That will be a great pity. (Cited in Krömer 2007: 152–3)

It is difficult to judge whether or to what extent this attack reflects a too one-sided view of the situation, both at the personal level but also with respect to Weil's opinions. Nevertheless, there can be no doubt that the very existence of a document of this kind, which circulated among the members of the group in 1960, bears clear witness to the intensity of the turmoil that took place at the time. The rising importance of new generalizations based on algebraic concepts—categories, functors and homological tools in the first instance, and then Grothendieck's increasingly abstract ideas—placed strains on the entire Bourbaki enterprise that accelerated the demise of this once ambitious and highly influential project.

FINAL REFLECTIONS

During the final decade of the twentieth century, the face of mathematics underwent deep changes. Given the new ways in which the invention of mathematical knowledge came to be widely practiced, the images of mathematics promoted by the Soviet school and the Bourbaki group gradually lose their vitality and centrality. Changes included the flourishing of new trends in the pure fields of mathematics that were less akin to the Bourbaki approach, a marked momentum in various fields of applied mathematics, and the massive adoption of computational techniques based on the use of digital devices.

The USSR was unable to sustain the technological and educational triumphs of the 1960s; and in the 1970s, a period of economic contraction began and came to be known in Russia as stagnation (*zastoi*). It was in large measure caused by entrenched, self-seeking bureaucrats who controlled economic life and managed it for their own benefit. The attempt to break out of this economic depression, with its chronic shortages of ordinary consumer goods such as laundry detergent, through more open public debate (*glasnost'*) and structural reforms (*perestroika*) led to increased social unrest. After an attempted coup by the old guard failed in the autumn of 1991, the complete breakup of the Soviet Union followed on rapidly. During all these decades, Soviet mathematics continued to be carried on at a high level, despite suffering damage in the form of the emigration of some of its best minds to Germany, Sweden, Britain, France, Israel, Canada, and the United States, a flow that increased sharply after the USSR supported the Arab side against Israel in the 1967 war. Altogether, some 1,000 mathematicians emigrated from the USSR just before and after it broke up.[10] This was a great hemorrhage of talent that one would think no country could afford.

Nevertheless, there were enough outstanding mathematicians in the country that the tradition of excellence continued, even during the 1990s, a very difficult period of shortages and inflation. In fact, a veritable nova of a star in the mathematical firmament burst forth at this time. The famous Poincaré conjecture, which asserts that a three-sphere is topologically determined by its homotopy groups, a conjecture that had been open for a century, was finally settled in 2006 by a young mathematician from Saint Petersburg, working in isolation. Grigori Ya. Perel'man, who had earlier established a solid record of mathematical achievement, published his result on the Internet. For this outstanding result, he was awarded the highest honor available to a mathematician, the Fields Prize, and was the only person ever to decline this honor. Having apparently no interest in either fame or fortune, he wished to be left alone to pursue his mathematical interests (see Chapter 4).

The Bourbaki project reached its high point of success and influence during the 1960s, when structuralist ideas in general were much in vogue. Directly or indirectly, the Bourbaki project also influenced fields like economics (via the work of Gerard Debreu), as well as psychology and anthropology (via Jean Piaget and Claude Lévi-Strauss). It also provided inspiration for the collective of experimental literary writing Oulipo (created in the 1960s by Raymond Queneau and François Le Lionnais). More infamously, critics of the "New Math," a reform movement launched during the 1960s in French elementary and high schools as well as in other countries, often pointed to Bourbaki as an inspirational force behind this ill-fated pedagogical approach (see Chapter 1). However, Jean Dieudonné emphatically denied any direct involvement of Bourbaki members in fostering this experiment.

By the 1980s, however, the impetus that had characterized the project in its early years could no longer be maintained. Dieudonné's catalyzing role could hardly be matched after his retirement. Some new chapters were proposed which never materialized, on topics such as analysis of several complex variables, homotopy theory, spectral theory of operators, and symplectic geometry. Nothing came either of plans to rewrite the first six books. The new books that did appear by 1980 included a summary of differential and analytic manifolds, seven chapters on commutative algebra, eight chapters on Lie groups and Lie algebras, and two chapters on spectral theories. In the 1970s the group found itself involved in a legal dispute with its publisher, which absorbed a great amount of energy.

The *Séminaire Nicolas Bourbaki* continues to operate to this day, and it has since become a major platform for showcasing contemporary mathematical invention. Along with other Parisian institutions, this *Séminaire* has gained prestige as many talented mathematicians have used it as a stage for presenting breakthrough results in a wide variety of fields. These usually have little or nothing to do with the initial Bourbaki project, of course, but the participants nevertheless share a sense of continuing the larger spiritual undertaking that began here after the Second World War.

As one would expect, certain modern developments were of vital importance to Soviet mathematicians as well as Bourbaki, for example, the idea of a vector space of functions, which lies at the heart of functional analysis. These common interests notwithstanding, though, it appears that Bourbaki never had many champions in the Soviet Union. Their approach faced strong headwinds coming from the direction of the Marxist worldview. In general, the ideas of Bourbaki were most relevant in establishing a common language of mathematical discourse (especially matters of notation, allowable principles of inference in mathematical proofs, and the interpretation of mathematical propositions), but also in recognizing the need to educate young mathematicians in such concepts. Thus, to the extent one can speak of influence or affinities, they are most likely to lie in the areas of methodology and pedagogy. Prominent research mathematicians in the USSR, from Luzin to Kolmogorov, took a keen interest in these matters, but it is difficult to trace any widespread penetration of the Bourbaki program into Soviet mathematics. The English version of the 1950 Bourbaki (Dieudonné) manifesto "The Architecture of Mathematics" appeared a decade later in Russian translation as part of a collection of articles devoted to mathematical education. Although it exerted no deep impact, one can trace a specific influence dating from that publication. From that time hence, one frequently read in Soviet literature that the subject matter of mathematics was not restricted to problems connected with space and numbers: modern mathematics was also rooted in the study of *structures* (*struktury*). Aside from the areas just mentioned, the Bourbaki approach was closest to Soviet research in the area of control theory, where the axiomatic method and mathematical structures are relevant.

One prominent Soviet mathematician's critical (and perhaps atypical) stance toward Bourbaki surfaced in the year 2000, when Vladimir I. Arnold debated Jean-Pierre Serre. This exchange took place after Serre had written to Arnold saying that mathematics and physics have "nothing in common." Arnold defended a point of view that became popular after the triumph of general relativity, claiming that (applied) geometry was part of physics. Interest in General Relativity, was widespread, from the groundbreaking papers in the 1920s by A. A. Friedmann and Georges Lemaître to the more recent work of mathematical physicists, such as Stephen Hawking, Roger Penrose, and the Soviet group led by Jakov Zel'dovich (Thorne 1994). Arnold believed that he had convinced most of the audience of his position, but that the mathematicians among them were unanimously supporting Serre because of the authoritarian nature of French academia.

One obvious reason for Bourbaki's limited success was a fundamental incompatibility between its lofty abstractions and the organizing principles of Soviet society based on the economic ideas of Marx, Engels, Lenin, and Stalin. Soviet political leaders naturally felt inclined to favor mathematics with potential to increase production or advance technology. One such area was computational methods, where mathematical logic and foundational questions play an important role. These areas of pure mathematics, however, required practical implementation in order to promote technological innovation. This meant developing high-speed computers, the province of electrical engineers, a need they unable to meet adequately. What mathematicians could do without such devices, namely, develop techniques of approximation and practical substitutes for abstract existence theorems, the Soviet school of constructivism largely managed to accomplish.

CHAPTER FOUR

Mathematics and Worldviews

Competing Visions in the Age of Ideologies

JOSÉ FERREIRÓS

The mathematical developments of the last century unfolded against the background of competing ideologies, some of which continue to have echoes up to the present day. In general terms, these rival views evolved over three periods: the era of the two World Wars, 1914 to 1945, the post-war years from 1945 to 1990, and the thirty years 1990 to 2020. Eric Hobsbawm called the "short century" from 1914 to 1991 the Age of Extremes (Hobsbawm 1995); it saw extraordinary confrontations between competing ideologies, forms of social organization, and worldviews. The period 1914–45 not only witnessed the Great Wars, they also led to struggles for a "new man" and a "new society," with fascism and communism offering competing visions. Moreover, this era marked the effective end of European colonial domination. It also witnessed the rise of mathematical modernism and the clash of worldviews manifested in the foundational debates. This conflict was both a power struggle within the discipline as well as part of an effort to define normative standards for mathematical research.

The period from 1945 to the fall of the Soviet Union in 1991 was an era of consolidation for the two great ideologies, the Democratic-Capitalist vs. the Partocratic-Communist. That rivalry unleashed the Cold War era, a time that also saw the emergence of numerous postcolonial countries. During this period, a strong consensus emerged among mathematicians in the West, associated with names such as Hilbert and Bourbaki. The research norms that came to fore reflected disciplinary autonomy alongside a unified conception of mathematical structures anchored by stringent conditions for mathematical proof. The purism of this trend had little influence on mathematical research in the Soviet Union, however, where such an ideal was antithetical to the state ideology of Marxism. Soviet mathematics was, in fact, highly eclectic, but its practitioners advocated socialist ideals and rejected bourgeois trends identified with the capitalist West.

The period 1990–2020 brought swift globalization and the rise of a neoliberal ideology that tended to undercut social-democratic measures in favor of opening new markets. Other trends led to critical and still ongoing debates over feminism, post-colonialism, ecologism, and postmodernism. After the fall of the Soviet Union, competing ideologies no longer played a leading role in the world of mathematics, which became highly diverse and increasingly interdisciplinary. On the one hand, this era saw several unanticipated conceptual developments that served to undermine the former foundational picture; on the other, the IT revolution exerted a profound influence on many spheres of mathematical research. If, throughout most of the last century, much mathematical work was largely invisible, in today's highly digitalized world many are well aware that their lives have been deeply affected by the products and byproducts of mathematical knowledge.

Mathematical results are not a matter of opinion, at least not within their relevant contexts or their theoretical niches. For instance, all experts agree that proving certain important theorems in analysis and set theory requires the Axiom of Choice (see Chapter 3). On the other hand, many statements about mathematics are clearly open to dispute. Opinions tend to differ greatly when discussing the nature of mathematical knowledge or the suitability of mathematization in different spheres of society. Some fundamental questions—whether one discovers mathematics or creates it, or whether it represents a priori or empirical knowledge—have long been debated. Johannes Kepler, for example, regarded mathematics as entirely a priori and in a sense God-given, but during the twentieth century many saw mathematics as a purely human conceptual creation. One may safely assume that different worldviews, with their respective value-systems, will lead to sharply differing visions of mathematics.

At the level of philosophy, mathematics poses the question of the role of abstract objects in human knowledge, an issue central for Platonism. People have often wondered about the 'nature' of numbers, not to mention abstract structures—what does it even mean to say that numbers exist? Where do they come from? Mathematics has long been a true haven of abstract objects, which many tend to interpret as immaterial or even supernatural, the position of Platonists. Those who have opposed this view have generally adopted a naturalist or materialist position. Around 1900, Georg Cantor held that numbers were as real as material objects, actually even *more* real, since they are Ideas in the mind of God. Cantor's philosophy might well bear comparison with the religious Platonism of Kepler, were it not for the fact that Cantor created (discovered?) a theory of *transfinite* numbers. Indeed, he and Richard Dedekind (who was less metaphysically inclined) were among the first mathematicians to develop theories based on infinite objects, leading to the modern theory of sets (Ferreirós [1999] 2007). Another remarkable case was Kurt Gödel, an extraordinarily influential logician who, around 1945, defended Platonism not only for ordinary numbers but also for the whole universe of sets. Many mathematicians are naturally inclined to accept that there is a body of *mathematical objects* and *facts about them*, and their relations, that is independent of and unaffected by human endeavors to discover them.[1] And so mathematics often figures prominently in attempts to defend some form of idealism, or belief in supernatural entities.

The links between worldviews and mathematics in the twentieth century form an extremely broad topic. The question can be fascinating, given that the century produced many different ideologies. There were various cultural forms of modernism, scientism, and postmodernism; historians have grappled with various manifestations of fascism and all kinds of nationalisms; and of course, they have confronted standard sociopolitical ideologies, in particular capitalism vs. socialism/communism. One may be tempted to interpret the last century in terms of a confrontation between modernism and conservatism, both of which evolved and took on new forms over the course of the period. Yet, the spectrum of possibilities is actually more complex. For instance, Nazism and fascism, which many would consider deeply conservative and anti-modern, are viewed by many historians as modernist developments that aimed to establish an alternative form of modernity. These movements, much like those on the left, were obsessed with the idea of bringing about not just a new social and political order but also a "new man." As such, they were revolutionary movements that romanticized the idea of Revolution, a standard feature of twentieth-century modernisms.

Up until the First World War, positivism posed a potent challenge to traditional conservatism. Its secular vision of a road to human progress through science even took on religious overtones for those who worshiped the ideals set forth by August Comte (see Chapter 4 of Volume 5). More common, though, were those who espoused modernist worldviews based on an optimistic idea of secular progress that proclaimed the inevitable triumph of science and reason over traditional religious worldviews. This ongoing dialectic, contrary to some expectations, continued throughout the course of the last century and beyond, but the dynamic shifted markedly, as scientism and positivism fell out of favor among the academic elite. Postwar tensions contributed to what some called an "Age of Anxiety," as the darker side of scientific progress became more evident—most notably in the form of atomic bombs, but also with chemical weapons, environmental pollution (beginning with the case of DDT in the 1960s), and other undesirable byproducts of scientific technology. This trend also affected mathematics. The artificial civilizations we live in today are created and sustained by engineering know-how and mathematical techniques. More than ever before, mathematics has become an integral part of our technological domination of nature.

By the end of the century, those who enjoyed the benefits of living in advanced economies had come to take many features of modern life for granted. Clearly, modernism has not ushered in religion-free societies that celebrate the latest scientific breakthrough or discovery. Instead, a resurgence of religious life took place, though often not in traditional forms but rather following new "alternative religions" or New Age ideas. In the broader sense, many of these twentieth-century movements reflected a skeptical, if not outright hostile view of scientific rationality that would have been far more difficult to find among educated people during the middle decades of the century. Today, science bashing (e.g., anti-vaccine movements) has become commonplace, though a more general disaffection for scientific rationality had long before laid the groundwork for the more radical manifestations felt today.

This essay will consider how deep shifts in worldviews affected the practices of mathematicians who lived in societies that underwent jolting change. A prominent example, already discussed in Chapter 3, is the case of the Soviet Union from the 1930s, when Stalin sought to impose Marxism as the official state ideology. In general terms, this inquiry concerns the question of how different worldviews (Weltanschauungen, мировоззрения, cosmovisiones, 世界观) have affected mathematics and its cultural perception. In many cases, mathematicians freely adopted a philosophy or worldview that reflected their own self-understanding of what it means to do mathematics. In such instances, the question arises as to how these larger views reflect societal norms. Even Platonism reflects a clear picture of the mathematician's role in society. So understood, worldviews are never merely descriptive in character, quite the contrary: they include normative aspects at their core, so that values and evaluation are central to them. They typically involve an idealized vision of society and the nature of humankind's relation to the world at large, including the meaning of life. Thus, two long-lasting changes in worldview, that rose to prominence in the twentieth century, but continue to act powerfully today, are feminism and ecologism.

MATHEMATICAL AND SCIENTIFIC KNOWLEDGE

One debate that runs throughout the whole twentieth century concerns the *autonomy* of *pure* mathematics, and the issue of the relationship between science and mathematics. We have to consider different perspectives on the relation between mathematics and the empirical world, but also the world of reason. By mid-century, the idea of a fully autonomous pure mathematics reigned supreme, but the turn of the twenty-first century has seen a return to an emphasis on the practical utility of mathematics.

Considering the central role that mathematics plays in the natural sciences and engineering, one can adopt one of two simple philosophical stances. The first attaches a deep metaphysical significance to mathematical ideas: Albert Einstein said that experience justifies the belief that nature "is the realization of the simplest conceivable mathematical ideas"; experience remains the criterion of the physical validity of theories, "but the creative principle resides in mathematics."[2] The Nobel laureate Paul Dirac (see Chapter 5) expressed a similar view:

> It seems to be one of the fundamental features of nature that fundamental physical laws are described in terms of a mathematical theory of great beauty and power, needing quite a high standard of mathematics for one to understand it. . . One could perhaps describe the situation by saying that God is a mathematician of a very high order, and He used very advanced mathematics in constructing the universe.[3]

The second stance insists on pure empiricism, which was a watchword for the influential philosopher-physicist Ernst Mach, a key figure for modern positivism. Mach rejected the use of models in physics (see Chapter 5), as for example atomic models, claiming that the appeal to unobservable physical constructs amounted to the importation of illegitimate metaphysical ideas. This standpoint clearly begged for an answer to the question: what kind of knowledge is mathematics? Mach's followers, the logical positivists (a new, modernist version of positivism), claimed this was logical knowledge, that is to say mathematics is "analytic" or "tautological," in the following sense. This proposal involves a clear basic scheme: there are only two sources of knowledge, experience and logic. All knowledge of real phenomena derives from observational data, while mathematics (e.g., the infinitesimal calculus) must be like the simplest forms of logic, a matter of mere linguistic reformulation.

This viewpoint was influential during the early and mid-twentieth century and was associated with the renowned Vienna Circle, a group of positivist philosophers and scientists who gathered in the capital of Austria. They proclaimed the triumph of a "scientific philosophy" and the demise of metaphysics, views laid out in a famous 1929 manifesto entitled *The Scientific Conception of the World. The Vienna Circle*. A key member of the group was the mathematician Hans Hahn, who stated that logic is "a prescription for saying the same thing in various ways and for extracting from what is said all that is (in a strict sense) connoted."[4] Until his death in 1934, Hahn contributed

FIGURE 4.1. Einstein bestowing the first Albert Einstein Award for Achievement in the Natural Sciences to Kurt Gödel and Julian Schwinger, March 14, 1951. Photo by Bettmann / Getty Images.

to the Vienna Circle as a leading participant in the discussions—he was, in fact, the principal signer of the Vienna Circle manifesto. He claimed that mathematics is made up of *tautologies*—statements that can only be true, in virtue of the meaning of the terms involved (classic examples are "A is A" or "if A then A"). Thus, mathematics is devoid of content, a conception laid out in the work of Ludwig Wittgenstein's celebrated treatise *Tractatus logico-philosophicus*. Hahn said that Wittgenstein's *Tractatus* "has explained the role of logic" and mathematics.

We understand by *logicism* the view that mathematics ultimately reduces to pure logic. Some years earlier, Henri Poincaré took up his pen to attack this kind of position. He wrote several pieces disputing the views of Bertrand Russell, another famous philosopher who collaborated with Wittgenstein, defended logicism as a deep discovery, and served as a model figure for members of the Vienna Circle. Russell was the author (with A. N. Whitehead) of a famous work that attempted to establish in full detail the logical foundations of mathematics, *Principia Mathematica*. Poincaré was astonished to see that influential thinkers defended a conception of mathematics as tautological and, hence, devoid of content. He insisted on the key role of *intuition* in mathematics, and some have called him a forerunner of modern *intuitionism*. Any discussion of logicists vs. intuitionists leads to complications due to two facts: the participants held quite different views on the scope and limits of logic, whereas the very idea of 'intuition' is itself highly ambiguous.

In any event, one can say that the conception of mathematics as "analytic" or "tautological," devoid of content, fell out of favor. Most experts in logic and foundations gradually abandoned this standpoint in the wake of contributions and analyses offered by Kurt Gödel and Alfred Tarski in the 1930s. Thereafter, philosophers returned to a more complex conception of mathematics and its role in science, adopting positions closer to ideas Poincaré and others had promoted earlier.

In view of this, Poincaré's conception offers a good standard against which to compare the images of mathematics during the twentieth century.[5] He regarded mathematics from an inclusive viewpoint and took into account a wide spectrum of its forms and manifestations. His basic position became famous and influential: according to Poincaré, logical rigor is a necessary form of hygiene, but creative power resides in intuition. Mathematics is not a matter of symbols, logic and formalization, but an expression of contents whose ultimate source is intuitive. The primary mathematical intuition is not geometric, however, but arithmetical—above all, the reasoning that proves properties for all numbers by considering the transition from n to $n+1$, i.e., the *principle of mathematical induction*.

Hence, the natural numbers are primary givens of mathematical knowledge, though many mathematical ideas are triggered by physical phenomena and sensory-motor experiences. These lead us to the conception of a system of three-dimensional geometry, which for reasons of simplicity we take to be Euclidean geometry. To understand movement and physical interactions we are led to conceive a geometrical universe, and a continuous time understood abstractly. By convention, geometry will take its simplest form—we find here one of the forms of Poincaré's *conventionalism*, but note that such conventions are not arbitrary (they are not merely name giving). The experience of movement will also suggest the idea that bodies and extended space are infinitely divisible (hence the need for fractions), and even more, the sophisticated idea of the continuum, especially the prototypical continuum of real numbers.

The process of formation of the idea of the continuum from more basic notions of number is not *a priori* or purely logical (on this, Poincaré thought the logicists were wrong). Mathematics in general is not *a priori*, its growth is always guided by natural science, or by experiences of the physical world, which suggest all kinds of problems and notions that, otherwise, the mathematician would never have imagined. Thus, we come to the idea of *function* or functional relation, which grew more and more complex and rich from the seventeenth to the twentieth century. We start with simple functions, say polynomials, but move on to trigonometric functions (sine, tangent, etc., discovered in the setting of astronomy) and on again to infinite series, to Fourier series, and beyond. It is physical experience and the imagination of phenomenological possibilities that take us to the most general concept of function and the idea of arbitrary functions.

Set theory can be understood simply as the product of an effort to *systematize* all of the above, reducing a series of systems (number systems, real functions, function spaces, and so on) to a uniform foundation. The notion of set, and the relevant axioms, make it possible to justify the various steps and open up ever-broader systems, without being too difficult to interpret or visualize. Poincaré had objections to Cantor's ideas about the transfinite sets, but we need not enter into this matter.

The task of the mathematician is to solve abstract problems, presented in symbolic form, or methodological and conceptual difficulties which emerge in the interrelation of concepts of number, function, algorithm, structure. Such a task can be developed without paying attention to the real situations that may have stimulated the posing of the problem. According to Poincaré, the mathematician as such is a *pure* mathematician—yet, mathematics is always tied to its applications. From this point of view, the so-called "problem of the applicability of mathematics" is a pseudo-problem engendered by a biased understanding of mathematics. The mathematician may know little about the application of his theories, and insofar his or her ability to elaborate mathematical models for empirical situations will be limited. This will limit their ability to cooperate with other scientists, but that is a problem of education and training, not a problem of mathematics per se.

A younger contemporary quite close to Poincaré's way of thinking was David Hilbert, the most famous twentieth-century mathematician. True, Hilbert was more inclined to a purist vision of mathematics, and he was somewhat less involved in physical pursuits than Poincaré. He also claimed that mathematical knowledge was objective and without presuppositions; it required neither a transcendental intelligence as claimed by theologians nor a special faculty of our understanding, as the intuitionists maintained. Yet, Hilbert essentially agreed with the inclusive viewpoint of his older French colleague, as can be seen from his famous lecture "Mathematical Problems" (1900, see Introduction) and his lectures on "Nature and Mathematical Knowledge." Although Hilbert advocated axiomatics, he also thought that geometry had clear empirical origins (Corry 2006). Thus, he argued that the gradual deepening of geometrical knowledge opened up the possibility of considering structural systems that had no physical correlate.

Hilbert's name is also invariably associated with *formalism*, which claims that mathematical theories are just a game with meaningless symbols (played according to precise pre-determined rules). This attribution, however, is misleading, as Hilbert was definitely not a strict formalist. In the foundations of mathematics, a *strict formalist* defends the position that mathematical theories are purely symbolic constructs with no particular subject matter. This means one can adopt as axioms any combination of symbols whatsoever, from which one derives consequences by carefully established (but purely formal or syntactic) rules. The theory is the totality of (in principle meaningless) 'sentences' that can be derived from the arbitrarily chosen axioms; with the only, but essential requirement that no contradiction appears in this "game of symbols." Hilbert employed the *methodology* of formalization in the context of his attempt to defend 'classical' mathematics by means of *proof theory* (see below), but his understanding of mathematics was much more nuanced and rich. It was not Hilbert, but a younger generation in the 1930s who reinterpreted his ideas and proposals, turning them into a reductionist philosophy of mathematics.

The impact of the new style of axiomatics, promoted forcefully by Hilbert's *Foundations of Geometry*, together with the strong independent status enjoyed by mathematics as a discipline, led to a characteristic twentieth century phenomenon: *modern mathematics*. The novelty of modern mathematics lies in its clear self-conception as an autonomous intellectual endeavor. Whatever may be the origins of basic concepts like number or figure (be they empirical phenomena, innate intuitive givens, or intellectual creations), the mathematician is free to combine them, modify them, postulate interrelations, and so on, giving rise to a rich world of mathematical *structures*. Such relational structures describe *in abstracto* possible, hypothetical 'states of affairs', and it is the task of the mathematician to investigate what consequences follow; if necessary, introducing novel methods to solve problems, or developing new structures to overcome obstacles that may emerge in the setting as previously given.

A pioneering figure for modern mathematics was Felix Hausdorff, who wrote a very influential textbook on set theory and point set topology. For Hausdorff, mathematics was fully autonomous, and he rejected any kind of heteronomy (dependence on metaphysics, or on the empirical world). Mathematicians were altogether free to explore all conceivable conceptual possibilities, any of which might prove to be legitimate if shown to be logically sound (Epple 2006). Hausdorff thus promoted a version of "Formalismus," which also entailed a new understanding of the notion of space. His idea was to treat this concept purely mathematically, which meant completely independent of all experience or any assumptions about the reality or ontological nature of abstract spaces. Hausdorff adopted the same position with respect to time, which from a mathematical standpoint might be continuous or discrete, linear or cyclic, etc. In any case, he advocated the study of space and time in the context of formal systems of principles, the axioms. These latter were freely conceived assumptions, from which the mathematician derived logical consequences. No form of apriorism was admissible; on the contrary, according to Hausdorff, the task of mathematics was to dissolve apparent certainties by showing the infinite complexity inherent in all possible systems. Guided by this modernist conception, he developed the standard set of axioms for general topological spaces and introduced the highly non-classical notion of Hausdorff dimension that underlies so-called fractal geometry.

ABSTRACT MODERNISM

The new conditions of intellectual life in the twentieth century strongly favored a purist value-system among Western mathematicians. Many of the seeds for that development had been sown decades earlier, particularly due to the influence of German institutions and educational innovations (see Chapter 2 of Volume 5). A central principle in the ethos of the modern research university called for the pursuit of science for science's sake with an attendant emphasis on *pure* mathematics. Yet, the strength of this trend in the twentieth-century went far beyond the walls of the universities. Looking backward from 1991, the distinguished mathematician and former Fields medalist David Mumford underscored how the impulse toward purism was nearly omnipresent:

> The 20th century has been, until recently, an era of 'modern mathematics' in a sense quite parallel to 'modern art' or 'modern architecture' or 'modern music'. That is to say, it turned to an analysis of abstraction, it glorified purity and tried to simplify its results until the roots of each idea were manifest. These trends started in the work of Hilbert in Germany, were greatly expanded in France by a secret mathematical club known as "Bourbaki," . . . (Gray 2008: 185)

In a similar vein, Jeremy Gray has emphasized the similarities between this scientific movement and its counterparts in the arts, law, and music. In all cases, it was "the de facto view of a coherent group of people, such as a professional, discipline-based group, who were profoundly serious in their intentions" (Gray 2006: 185). Core examples include cubism in painting, serialism in music, and the abandonment of narrative in literature.

There were two sides to these developments, one conceptual and the other social. Several fields of mathematical research underwent a sharp shift in orientation. Algebra, once dominated by the theory of equations and methods for solving them, now focused on the study of abstract algebraic structures (Corry 2004). Analysis was long the province of concretely defined and generally well-behaved functions, but modern measure theories opened a vast new realm of possibilities for studying entire new classes of functions. Euclidean geometry lost its paradigmatic status over the course of the late nineteenth century, and the trend toward axiomatization and abstraction created a proliferation of new geometries as well as a broad new understanding of the notion of a mathematical space.

Turning to the social dimensions of this shift toward modernism, the landscape of mathematical communities also underwent profound changes. By the twentieth century, well-established professional communities of mathematicians had emerged with strong institutional support. This led to a reorientation of the discipline, which increasingly freed itself from the study of natural phenomena. By severing its former ties with astronomy and physics, mathematicians began to enjoy full autonomy in the development of their ideas and methods. Independent journals, run only by colleague mathematicians and devoted to pure mathematics, ensure publication; university departments are well established and unquestioned, expanding in places like the USA; the community can work entirely on its own. The result will be free 'modern' mathematics, an emphasis on absolute purity and abstractness, on the formal aspects of the work (axiomatics, structures, logic), and even—in some quarters—a measure of disrespect for the so-called applications.[6]

These developments reached a kind of peak with Hilbert's attempt to make mathematics fully self-sufficient by showing, in principle, how it could account for its own foundations. His approach, later called *metamathematics* or *proof theory*, studied mathematical theories "from above," as it were, by making the proofs themselves the objects of mathematical study. Hilbert and his colleagues wanted to deepen the foundations of mathematics by establishing more precise axiom systems while *formalizing* them with the methods of modern logic, hence the name "formalism." Having perfected the calculus of logic, and having used it to establish the axioms, the next step was to study the statements provable from them. This provided a method for studying mathematical theories by analyzing their proof structures, including plans for a proof establishing the *consistency* of the theory itself (such a proof would show that one could never derive a contradiction within the theory).

Hilbert's proof theory arose in answer to the famous "foundational crisis" in mathematics, which had much to do with the methods of set theory. The modern mathematics of sets and structures rested on principles (like the Axiom of Choice, see Chapter 3) that made many experts uncomfortable. Some considered this a dubious principle; others rejected it altogether. Moreover, a naïve approach to sets led to logical contradictions, like the celebrated Russell Paradox. As a result, some mathematicians began proposing alternative methods for founding and developing mathematics, a trend that gained momentum during the course of the First World War. Amsterdam's Luitzen E. J. Brouwer promoted an "intuitionistic set theory," which became the basis for so-called *intuitionistic* mathematics. Brouwer's approach represented a radical departure from classical mathematics, insofar as it rejected the standard laws of logic when applied to infinite sets. In particular, he claimed that it was illegitimate to claim that a property of an infinite set is a priori either true or false. Hilbert initially ignored Brouwer's challenge, but his former student, Hermann Weyl, met with Brouwer, who converted him to intuitionism. Weyl had already been exploring alternatives to standard set theory, but he now gave these up and in 1921 published "On the new Foundational Crisis in Mathematics." In dramatic language, he threw himself behind Brouwer's cause and declared: "Brouwer . . . that is the Revolution!"[7]

In responding, Hilbert fired off some explosive shots in defense of modern methods and Cantorian set theory.[8] This was no *revolution*, he assured his fellow mathematicians, but only an attempted *putsch* that aimed to impose dictatorial restrictions on mathematical reasoning—and it was doomed to failure. Brouwer against Hilbert, the mathematics of intuitive certainty against the abstract conception of freely postulated structures; the critic of transfinite set theory against the defender of Cantor's "paradise"; the extravagant thinker (but outstanding mathematician) who demanded special logical forms of proof when working with infinity, against the 'classical' defender of *reductio ad absurdum* and purely existential proofs. The story could hardly fail to capture the imagination, and it has indeed become one of the most famous conflicts in the history of mathematics.

In the ensuing debate, the status of Cantor's theory of transfinite numbers became an emblematic matter of contention. Hilbert's 1925 paper "On the Infinite" featured an attempt to prove the Continuum Hypothesis (CH), which Brouwer and other constructivists regarded as a pseudo-problem. Although the proof he sketched failed, advances in logic eventually led to a breakthrough in the 1960s: like the axiom of choice (AC), mathematicians now know that CH is independent of the other standard axioms for set theory. Hilbert's paper also described his larger goal of salvaging the fruitful definitions and deductive methods Cantor had introduced. Convinced of their soundness and intrinsic worth, he wanted to nurse and strengthen them. In a barb clearly directed at Brouwer, he proclaimed, "No one shall drive us out of the paradise which Cantor has created for us" (van Dalen 2013: 455). This appeal thus aimed to quash all criticism of modern structural mathematics by relying on proof theory to establish "the same certitude for our deductions as exists in ordinary elementary number theory, which no one doubts." Meanwhile, Brouwer worked away meticulously on his intuitionistic approach to foundational problems, whereas Weyl withdrew from the battlefield to work in less controversial corners of mathematics.

The historian Herbert Mehrtens argued that this foundational debate was more than just a matter of abstract principles and technical means of proof.[9] He saw the fight between Hilbert and Brouwer as part of an effort by German nationalists to gain control of the discipline during a particularly contentious political period. In some respects, this power struggle within the German mathematical community bears comparison with the one that took place later in Moscow, where younger Soviet mathematicians helped in toppling Nikolai N. Luzin from power (see below and Chapter 3). Brouwer had no realistic chance of displacing Hilbert, but he formed an alliance with Berlin's Ludwig Bieberbach, who would later become infamous as a leading Nazi mathematician and editor of the journal "German Mathematics" (*Deutsche Mathematik*). Although decidedly pro-German, Brouwer was no anti-Semite, so he and Bieberbach finally went their separate ways in the 1930s.

Hitler's Nazi Party borrowed some of the rhetoric and pageantry that Mussolini made famous in Italy (on mathematics under Italian fascism, see (Guerraggio and Nastasi 2005). Yet, these two variants of fascism differed in one crucial respect: Nazism was fundamentally a racist ideology driven by a deep resentment of foreign enemies outside the country and the Jewish minority within it. Judaism, in the eyes of Hitler's followers, was not a legitimate religion but rather signified those who belonged to a subversive ethnic group that threatened the body politic of the German people. Once in power, the Nazis took immediate steps to identify and remove Jews who held civil service positions, which included most who taught at German universities. In mathematics alone, this policy sparked an enormous brain drain: see Chapter 2 and Siegmund-Schultze (2009). Enrollments at the German universities fell dramatically during the Nazi period, so that by 1939 the number of students in all subjects was well below half of the high reached in 1932. The drop in mathematics was truly catastrophic: the numbers fell from over 4,000 in 1932 to around 300 in 1939! Biology, on the other hand, but in particular "racial science," enjoyed considerable popularity under the Nazis. Within German mathematics, ideologues like Bieberbach played a marginal role, whereas the mainstream figures quickly fell in line and accommodated themselves to the anti-Semitic policies of the Nazi government (Segal 2003).

The earlier alliance between Bieberbach and Brouwer had more to do with academic politics than with mathematics. After the First World War, the victors refused to allow German scientists to participate at international conferences, including the international congresses held in Strasbourg in 1920 and in Toronto four years later. Brouwer considered this deeply unfair, and so he demanded that the Union Mathématique Internationale (UMI) lift this policy of exclusion. During preparations for the 1928 Congress held in Bologna, the German mathematical community learned that the Italians would officially invite them without seeking to amend the UMI statutes. In response, Brouwer and Bieberbach organized a counter-boycott of the ICM in Bologna, noting that the UMI boycott policy was still in place. Their efforts greatly annoyed Hilbert, a staunch internationalist who strongly supported this opportunity to restore cooperative relations.

These political tensions also spilled over into similar difficulties within the editorial board of *Mathematische Annalen*. Hilbert headed this prestigious journal, but he was old and seriously ill. Brouwer and Bieberbach, on the other hand, were two of the most active members on the board. Moreover, their political orientation resonated with the views of some more conservative board members. Just as the *Annalen*'s editorial board was politically divided, with Hilbert heading the liberal faction, so was the German mathematical community at large. The traditional rivalry between Göttingen and Berlin took on strong political overtones after 1920, when Richard Courant became director of the Göttingen Institute. Courant had been one of Hilbert's many Jewish students, and eventually he and his Jewish colleagues became targets of anti-Semitic resentment. These hostilities were perhaps only rarely articulated, but Hilbert was long aware of them. These circumstances, coupled with Brouwer's growing ambition, prompted Hilbert to expel the Dutchman from the editorial board (despite the opposition of two allies, Einstein and Carathéodory). For Brouwer, this came as a crushing blow, from which he never recovered (van Dalen 2013).

Brouwer's intuitionist methodology never really caught on with working mathematicians, even if it did contribute significantly to "constructivist" approaches. For someone like Ludwig Bieberbach, who had no serious interest in the foundations of mathematics, intuitionism had a broader, vaguely political appeal, since it rejected formalism and promoted values like "intuition" and "truth." After Hitler came to power, Bieberbach quickly donned a brown shirt and began his own campaign as the foremost spokesman for an ideologically motivated "Deutsche Mathematik" based on a pseudo-scientific theory of racial styles of thought in mathematics. His pronouncements provoked a mixture of outrage and disbelief among mathematicians abroad, and conservative supporters of the Nazi regime largely distanced themselves from Bieberbach and his entourage, who were not invited to take part at the Oslo ICM held in 1936 (Hollings *and* Siegmund-Schultze 2020).

In the end, Brouwer lost both the political and the foundational battle, despite the powerful negative consequences of Gödel's results for Hilbert's formalist program. Intuitionism continued to interest specialists who studied the implications of intuitionistic logic and methods, but it held very little appeal for working mathematicians, few of whom worried about philosophical problems like the meaning of mathematical existence. Brouwer's approach was simply too complicated, whereas the partisans of formalism and axiomatics stood on the side of productivity, efficiency, and economy of thought. By the late 1920s, when the political atmosphere grew calmer, few still claimed that mathematics was undergoing a crisis. On the contrary, most recognized this as a time of great expansion and intellectual achievement.

Ironically, a major triumph of "formalism" came at the expense of Hilbert's original foundational goals. His metamathematical project aimed to provide absolute consistency proofs for arithmetic and real analysis. Had he succeeded, he would have defeated all skeptical doubters; mathematics would have become entirely self-sufficient from the logical point of view. Yet, Hilbert's proof theory led to just the opposite result. It was in this context that the brilliant young Viennese mathematician, Kurt Gödel, proved two profoundly significant theorems. The first states that any formal axiomatic system rich enough to contain ordinary arithmetic must be either incomplete or inconsistent. The upshot was that no matter how many axioms one added to the system, it would always contain statements that were unprovable within the system, unless of course the system turned out to be inconsistent. In that case, though, the axioms lead to contradictions—every statement and its opposite would be provable. Gödel's second theorem went to the heart of Hilbert's program. Metamathematics made it possible to encode the consistency of the axiom system into a formal sentence (intuitively saying, "this system is consistent"), and Gödel proved that precisely *that* sentence could never be derived by the methods of Hilbert. Thus, Gödel's second theorem made it abundantly clear that Hilbert's original goal was unattainable, and later results in proof theory have not changed this situation.

None of these developments proved a hindrance for abstract modernism, however, as the momentum behind 'formalism' in the broad sense of structural freedom won the day. Gödel's results from 1931 appeared at the very same time that Bartel Leendert van der Waerden published the two volumes of his *Moderne Algebra*, based on lecture courses taught by Emil Artin and Emmy Noether. As described in Chapter 3, this textbook was the single most famous embodiment of modern mathematics in the interwar period. Abstract modernism, as a movement in mathematics, was not only an expression of the relevance of "forms" or structures, it also signaled the arrival of an era of solid institutional autonomy and intellectual independence. The institutional configuration of the universities and the organization of scientific research in general helped to mold a community of researchers united by an ideology of pure mathematics.

The foregoing account of the foundational crisis in mathematics and its passing underscores the relevance of contemporary sociopolitical issues, but this does not mean to imply that the intellectual dimension in this conflict was any less important (Ferreirós 2008; Gray 2008). No reduction of the foundational debates to purely sociopolitical "causes" is intended, only the acknowledgment that what transpired was more than a matter of internal issues and questions about methodology and logic. Both factors played an important role: there were issues of control of the discipline and the community, but also serious conceptual and methodological questions, studied and clarified through logic and analysis.

MODERN MATHEMATICS IN THE SOVIET STYLE

In the Soviet Union, different types of "capitalist bourgeois pseudoscience"—genetics, cybernetics, sociology, semiotics, or historical linguistics—were explicitly rejected, especially during Stalin's administration.[10] For this reason, one might have expected that Stalin's USSR would have led to an ideological form of mathematics shaped by the writings of Marx and Lenin. At the very least, the materialist worldview might have compelled Soviet mathematicians to restrict their research to topics directly linked with familiar physical or economic processes. Yet, nothing like that happened; on the contrary, a powerful, but also eclectic Russian school of mathematics emerged, though not without some birth pangs (see Chapter 3). How modern mathematics became compatible with Marxist-Leninism makes for a dramatic story.

The more radical versions of modernism—such as strict formalism with its insistence on the arbitrary nature of axioms—could hardly have found a welcome reception in the Soviet Union. This kind of approach was clearly antithetical to the materialist worldview, as espoused by Friedrich Engels when he asserted that pure mathematics deals with the space forms and quantity relations of the real world. In fact, all three of the standpoints adopted in the West during the foundations crisis—logicism, formalism, but also intuitionism—met strong opposition among the Soviets (see Chapter 3). They took these to be bourgeois forms of "false consciousness," which promoted an "idealistic" conception of mathematics as purely intellectual and fundamentally disconnected from the material world. Hence, these more recent philosophical positions were clearly incompatible with dialectical materialism in a way that Poincaré's older views were not.

As discussed in Chapter 3, Dmitri F. Egorov and Nikolai N. Luzin, the two leaders of the Moscow school of mathematics, cultivated ties with the West, particularly with mathematicians in Paris. Both also had strong religious leanings and befriended members of the clergy, including the theologian Pavel Florensky. During the relatively open atmosphere of the 1920s, all three of these friends tried to accommodate themselves to life in Soviet Russia. Over time, however, Egorov and Florensky came under heavy attack for their forthright religious lifestyles.[11] Various experiments in new forms of communism took place during that decade, and for much of it the Moscow school of Egorov and Luzin flourished quite apart from these revolutionary developments. They and their students continued to interact with mathematicians in Western Europe, and in some cases, this even led to noteworthy collaborations. Pavel Aleksandrov spent nearly every summer in Göttingen, where he occasionally taught topology courses. He reciprocated by inviting Emmy Noether to teach algebra in Moscow. On one of his visits to Göttingen, Aleksandrov met Heinz Hopf, an encounter that sparked a highly fruitful collaboration.

FIGURE 4.2. Misha L. Gromov lecturing at the Oberwolfach Conference "Differential Geometry, Mathematical Physics, Mathematics and Society," in celebration of the 60th birthday of Jean-Pierre Bourguignon, 2007. Oberwolfach Photo Collection. Gromov, a product of the Soviet School, was to leave for the U.S., then France, before the collapse of the USSR.

Relatively few mathematicians were Marxists, but Lenin was determined to break the power of the Russian Orthodox Church, and the Bolsheviks wasted no time in persecuting clerics and their flock. This situation intensified dramatically in 1929 when Stalin unleashed a cultural revolution that hunted down "idealists" and foes of orthodox Marxism. Luzin had been far more cautious than Egorov, and yet he, too, became a political target after 1930, when his enemies labeled him as a decadent bourgeois thinker. A key figure behind the effort to expose the leaders of the Moscow school was the Czech-born Marxist philosopher of science, Ernst Kolman, who was determined to root out all forms of opposition to the Marxist-Leninist worldview (see Graham 1993; Vucinich 2002; and Graham and Kantor 2009). As Stalin began to consolidate his power, the fate of mathematics in Moscow hung in the balance.

During the 1920s, Egorov was one of the most distinguished scientists in the Soviet Union. Not only was he director of the Institute for Mechanics and Mathematics (MechMath) at Moscow State University, he also served as president of the Moscow Mathematical Society. By the end of the decade, though, his support had largely eroded, in large part because of his stance against repression of the Russian Orthodox Church. Following mass arrests of Egorov's co-religionists, Kolman stepped up his attacks, calling him a "saboteur" who corrupted the youth. In the spring of 1930, authorities dismissed Egorov from his post as director of MechMath in favor of Otto Shmidt, a staunch Marxist. Later that year, they arrested Egorov and expelled him from the Moscow Mathematical Society. During his imprisonment, Egorov continued to practice his daily religious rituals, which provoked bullying by guards. In protest, he began a hunger strike, which led to his death (Graham and Kantor 2009: 134–8).

This turn of events left Luzin badly shaken, and he soon resigned from the Moscow Mathematical Society and no longer taught at Moscow State University. Nevertheless, as a full member of the Soviet Academy, he remained a powerful figure within the scientific establishment. As described in Chapter 3, many of Luzin's now eminent former students felt disgruntled by his high-handed manner. As a sign of the times, in 1931 the new Communist Academy published the proceedings of a symposium on the relationship of mathematics to dialectical materialism. Its unsigned introductory chapter presented a stinging critique of the Moscow school, especially the views of Luzin.

By this time, the state's cultural-political strategy had taken on clear form. Ernst Kolman and other Marxist zealots wanted nothing more than to remove the older generation of bourgeois scientists from positions of power—especially those who had once criticized the new ideology—and replace them by young experts who would be loyal to the Party. Luzin fully recognized the dangers and tried to avert them by such activities as visits to proletarian schools. But Kolman put a reporter from *Izvestiia* on his trail, and this led to an article by Luzin that exaggerated the abilities of the pupils at the school he had visited. Kolman orchestrated this plan with Lev Zakharovich Mekhlis, the editor of *Pravda*, which soon thereafter printed Kolman's vicious attack on Luzin in an article entitled "Our Enemies hiding behind a Soviet Mask." Mekhlis then wrote to the Central Committee of the Communist Party, requesting an investigation, which Stalin approved. Kolman's personal ambitions would later prove his own undoing (he spent six years in prison), but in this instance his machinations touched off the infamous "Luzin affair" in July 1936 (on the various factions involved, see Chapter 3).

Personal motives played an important role in this complicated situation, but perhaps even more the general atmosphere of fear on the very brink of the Great Purge that came immediately afterward. Aleksandrov and Kolmogorov both testified against their former teacher, but they were in a particularly vulnerable position because of their homosexual relationship, a crime in Stalin's USSR (see Graham and Kantor 2009: 184–6). Personal information of that sort was always useful in a system that relied on coercion and threats, or a combination of pressure and privileges, to maintain control. Graham and Kantor report that, shortly before his death, Kolmogorov stated that he would "fear 'them' [the secret police] to his last day." Some of the charges Luzin faced, such as "servility toward foreign science," were less serious. Aleksandrov testified that by publishing his work abroad Luzin had shown insufficient ideological awareness, though he and Kolmogorov had published important work in Germany.[12] After the "Luzin affair," though, Russian mathematicians stopped publishing abroad altogether.

Far more serious were the charges of "sabotage" and "enmity toward anything Soviet," which Kolman and others brought against Luzin. Such accusations, had they led to a conviction, might have even ended Luzin's life. Yet then came another stunning surprise, when the presiding officer at the hearings announced that he was dropping the more serious charges against Luzin, and so his ordeal ended with a reprimand and the closing of Luzin's department in the Steklov Institute. He remained, however, a member of the Academy, an honor Aleksandrov would only receive after Luzin's death. What had happened? In all likelihood, Stalin himself, counseled by the Nobel laureate Peter Kapitza, decided to intervene on behalf of Luzin. Many others were much less lucky; Luzin's close friend, Pavel Florensky died soon thereafter in a concentration camp, one of the many innocent victims from this dark period of Soviet history.

By this time, the relatively open atmosphere of the 1920s had disappeared altogether, leading to a difficult period marked by ideological adjustments and power struggles. Nevertheless, the "young professors" managed to save modern mathematics, despite all the pressure. No one in Moscow dared honor Egorov's name (he was buried in an unmarked grave in Kazan, located, ironically enough, near that of the famous Russian mathematician Nikolai Lobachevsky). The once adored Nikolai N. Luzin stood condemned for having engaged in activities unworthy of a Soviet scientist. Now without influence, his students, especially Kolmogorov and Aleksandrov, were Moscow's rising stars. Kolmogorov was faithful to communist ideas, but he also cleverly found ways to justify pure mathematics as part of the modern methods Soviet mathematicians employed in line with Marxist-Leninist principles.

Particularly important in this respect were Kolmogorov's contributions to the Great Soviet Encyclopedia under the general editorship of Otto Shmidt. Starting in 1938, he contributed the general entry on 'Mathematics' to all three editions. He adopted Engels' definition—mathematics is the science of "quantitative relations and spatial forms of the real world"—but with a new twist. Mathematical developments arose out of "the most elementary needs of economic life," and both spheres of human activity conditioned one another throughout the course of history. Hence, the higher forms of abstraction found in modern mathematical theories in no way represent a separation from reality. In effect, the stock of relations and forms under study in mathematics remains inseparably linked with the demands made by technology and the natural sciences. Kolmogorov went on to sketch a history of mathematics along these lines. He noted how, by the nineteenth century, new developments sometimes emerged from intrinsic problems posed within mathematics itself. Nevertheless, such studies eventually led to fruitful applications, just as new technological developments provided fresh stimuli for mathematical theorists.

This dialectical understanding of mathematics was clearly much closer to Poincaré's position than to the foundational views of Western theorists. Graham (1993: 118–19) remarks that Kolmogorov's line was also very different from the articles in the *Encyclopedia Britannica* written, around the same time, by Frank P. Ramsey and Alfred N. Whitehead (co-author of the famous *Principia Mathematica* with Bertrand Russell). Both of the latter were logicists, and Whitehead went so far as to say that the act of counting is irrelevant to the idea of number—a perfect example of what the Soviets would call "bourgeois idealism." Kolmogorov's views, on the other hand, were consistent with Marxism-Leninism, which asserted that the source of human knowledge derived from actions in the material world.

Kolmogorov also emphasized the fundamental unity of mathematical knowledge, a central tenant of modern mathematics in the Western tradition as well (see Chapter 3). He thus noted that "new relations emerge among its parts, sometimes in unusual or unexpected ways," for which reason the mathematical community should be on the lookout for such connections. Kolmogorov considered himself a pure mathematician, but throughout his career he insisted that many problems and key concepts "have arisen, by necessity, from the consideration of perfectly concrete physical problems" (*Foundations of Probability*, 1933: v, thus written before the Luzin affair). As he said years later:

> I admire those mathematicians who become important in technology, I absolutely acknowledge the relevance of computers and cybernetics for the future of mankind, yet I feel that pure mathematics, in its traditional form, has not handed over the place of honor that it obtained among the sciences. The only thing that could wipe it out would be a sharp division of mathematicians into two trends – those who cultivate the new abstract fields without considering their links with the real world that beget them, and those who busy themselves with 'applications' neglecting the need for a deep analysis of the theoretical basis. For this reason, I want to underscore the value and merit of the position of those mathematicians who understand the place and role of their science in the development of science and technology, and in all of human culture, but who patiently continue to develop pure mathematics as it corresponds to the inner logic of its evolution.[13]

Thus, Kolmogorov's insistence on unity emphasized the compatibility between pure and applied mathematics.

Soviet mathematics continued to thrive as a glowing example of success among the scientific disciplines. It managed to maintain a balance between theory and applications, while incorporating the 'modern' styles of Hilbert and Noether. The contrast between this story and the disastrous case of Lysekoism in biology and agronomy (see Graham 1993: appendix A and B) could hardly be greater. This does not mean, of course, that mathematicians confronted no obstacles. A striking case in point was the new theory of cybernetics, which dogmatists at first denounced as a "pseudo-science" and a tool of Western imperialism. Later (after Stalin), leading mathematicians embraced it with enormous enthusiasm, perhaps more than any other development in Soviet science. Thus, Sergei P. Sobolev and Alexei A. Liapunov heralded the full acceptance of cybernetics as the beginning of a new era in Soviet science, a viewpoint Kolmogorov considered quite exaggerated (see Vucinich 2002: 16–21).

PLURALIST TRENDS IN RECENT TIMES

The last third of the twentieth century featured significant innovations in mathematics, both conceptually and practically, enhanced by the growing importance of computers. Around 1990, mathematicians in the United States expressed worries about declining numbers of undergraduate and graduate students, but this trend soon changed. Nowadays mathematics attracts many students throughout the world, and graduates in mathematics enjoy high employment rates and command high salaries. Where they work and what they do, on the other hand, reflect how dramatically external conditions have changed. At the high end of research, pure mathematics has continued to grow, while at the same time a noticeable shift took place in the balance between applied and pure interests. Today, a more utilitarian attitude tends to prevail—part of the "return of Fourier" (see Chapter 3 of Volume 5). These trends, along with the many applications of mathematics in modeling, numerical methods, and computational algorithms, have led to a renewed perception of mathematics as a fundamental science. Public perception of the importance of mathematics has undoubtedly changed markedly during the Digital Age, shaped by familiarity with computers, smartphones, big data, and the impact of financial mathematics. Cultural philosophers, on the other hand, have often cast a critical eye on what many take to be the social benefits of scientific progress.

Followers of the philosopher Martin Heidegger tend to be critical, if not openly hostile to mathematics. They consider mathematics to be a predetermined Procrustean bed, into which one forces real phenomena. This "pre-figuration of beings in the world" in effect transforms such phenomena into objects of technological control. Heidegger's influence has been prominent, though he had little to say about mathematics per se. His approach to mathematics falls in line with a general perspective on science, which regards it not as primary but as *subsidiary* to technology; science and mathematics are thus mere tools for the technical domination of nature. Heidegger wrote in grand style that the fundamental feature of modern science is that it "is mathematical," claiming further that this mathematical perspective laid things out in order to exploit them:

> The mathematical is this fundamental position we take toward things by which we take up things as already given to us, and as they must and should be given. – It is the fundamental presupposition of the knowledge of things. . . . The *mathemata* are the things insofar as we take cognizance of them as what we already know them to be in advance, the body as the bodily, . . . the animal-like of the animal, the thingness of the thing, and so on. This genuine learning is therefore an extremely peculiar taking, a taking where one who only takes what one basically already has.[14]

Heidegger tends to mistake numbers for mathematics in general, and believes that mathematization is merely quantification—an oversimplified picture. Yet, the key idea that mathematizing is a basic way of conforming natural phenomena for technical domination offers an interesting metaphysical perspective.

The political philosopher Jürgen Habermas, an influential thinker in the postwar Frankfurt School and a fierce opponent of the metaphysical views of Heidegger, nevertheless voiced strongly critical opinions with regard to the social impact of mathematics. For Habermas, mathematics shares responsibility for many features of global capitalism, including the promotion of neoliberal economics, technological disruptions, oppressive financial systems, and the spread of military weapons. The role of mathematical modeling has become a familiar theme in the twenty-first century. This feature of economic life garnered considerable attention in the wake of the financial crisis of 2008, as described by the mathematician Ian Stewart:

> Was an equation to blame for the financial crash, then? Yes and no. [The] Black-Scholes [equation] may have contributed to the crash, but only because it was abused. In any case, the equation was just one ingredient in a rich stew of financial irresponsibility, political ineptitude, perverse incentives and lax regulation. Despite its supposed expertise, the financial sector performs no better than random guesswork. The stock market has spent 20 years going nowhere. The system is too complex to be run on error-strewn hunches and gut feelings, but current mathematical models don't represent reality adequately.[15]

It should come as no great surprise when philosophers and social critics take aim at the mathematical world, but some of the most celebrated research mathematicians of recent times have echoed similar criticisms. A number of such voices cried out during the protests against the Vietnam War (1955–75), a conflict that pitted a poorly armed country against a technological giant that tried to bomb the enemy into submission. Political and moral protests have long since attracted prominent mathematicians, who in some instances have drawn unwanted attention to themselves.

The cases of Alexander Grothendieck and Grigorii Ya. Perel'man, two of the most famous mathematicians of the last half century, come immediately to mind. In 1970, Grothendieck decided to abandon his prominent position at the *Institut des Hautes Études Scientifiques* (IHES) in order to protest its reliance on military funding. He afterward took up a religious life before withdrawing from modern society altogether. In *Reapings and Sowings* (1986), he portrays the mathematical community as governed by competition and status. Two years later, he turned down the Crafoord Prize and sent an open letter to the media, in which he criticized what he regarded as the declining ethics of the scientific community. The letter also expressed his belief that before the end of the century "totally unforeseen events" would lead to an "unprecedented collapse of civilization."[16]

Grigorii Ya. Perel'man, who startled the mathematical world with his ingenious proof of the Poincaré conjecture, shocked that same world a second time in 2010 when he refused to accept his reward for solving one of the Clay Institute's seven millennial problems: one million dollars. Some years before this, he turned down the Fields Medal, and reportedly explained this by saying: "I'm not interested in money or fame," and "I don't want to be on display like an animal in a zoo. I'm not a hero of mathematics."[17] Much like Grothendieck, Perel'man disliked the "star system" that has long typified seminal scientific achievements. He disagreed, in particular, with the way credit was allocated to someone like himself without giving due recognition to other key players.

To appreciate the contrast between such recent attitudes and the lofty sentiments typically expressed by leading mathematicians from the early decades of the twentieth century, consider these words from Princeton's Oswald Veblen: "Mathematics is one of the essential emanations of the human spirit, a thing to be valued in and for itself, like art or poetry."[18] This widely shared *topos* evokes a very different image of the mathematician, though one might still wonder whether ambition and competitive drive are just as prominent in mathematics as for those who achieve success in other spheres, such as arts and letters. Veblen's notion of the mathematician as artist touches on the aesthetic sensibilities that many attach to abstract ideas. One can often hear talk of a "beautiful idea" or a "wonderful proof," but also remarks disparaging a mathematical argument as ugly. So long as these assessments take place within

FIGURE 4.3. J. Robert Oppenheimer (*left*) and John von Neumann, June 10, 1952, standing in front of the computer built for the Atomic Energy Commission and operated at Princeton's Institute for Advanced Study. Photo by Bettmann / Getty Images.

the realm of esoteric pure mathematics, they appear quite meaningless to an outside observer. That impression quickly changes, though, when the mathematics carries technological implications. A dramatic example was the Manhattan Project and the part in it played by John von Neumann, one of several brilliant Eastern European refugee mathematicians who helped build the first nuclear weapons. The physicist J. Robert Oppenheimer, the first director of the atomic bomb project in Los Alamos, later made the telling remark: "When you see something that is technically sweet, you go ahead and do it."[19]

Whatever one might believe about the nature of mathematics, it is difficult to dispute the linkages between viewpoints on mathematics and the general philosophical worldviews current at a given time. Around 1900, in Europe, there was a widespread belief that mathematical reasoning stood on the solid foundation of a single logical system, namely classical logic. Gottlob Frege believed that, but so did most of his contemporaries, including Hilbert and Poincaré. A century later, following a proliferation of new logical systems and an entirely new understanding of mathematical logic, no reasonably educated mathematician still held this absolutist view. In fact, even set theory was no longer in vogue, as influential younger mathematicians began to take note of its weakness. Bill Thurston referred to its principles as "polite fictions," so he and others began considering alternative views, thereby promoting a pluralistic outlook.

This shift fits a more general pattern of cultural re-conception. The 'grand narrative' of mathematics coming to life in the 'Greek miracle' of Thales to Euclid—developing along meanders, but essentially forming one inevitable grand current, all the way up to the ideas of Hilbert—was a dominant viewpoint around 1900. By 2000, this picture appears simplistic and Eurocentric; today, the roots of mathematics in different cultures are being reclaimed and the contributions of Chinese, Indian, Arabic mathematics, etc. vindicated and carefully studied (Katz 2007). Historians have grown wary of grand or linear narratives. In recent years, they have devoted more effort to the exploration of changing conceptions, blind alleys, and local situations. Running parallel with this trend, one finds many present-day proposals that reject the notion of a unique monolithic framework for understanding mathematics. This older modernist view has given way to a pluralistic outlook that makes room for many possible frameworks. A broader and more dynamic picture of mathematical activity has since emerged that stresses the processes that lead to change rather than the static features of mathematical knowledge. So understood, mathematics develops structural conceptions, but these are not rigid, because mathematical thinking constitutes an exploration of the many possible structural "states of affairs," their interrelations, and possible variations.

A century ago, the world was structured around colonial domination, while at the same time, the language and ideas of nineteenth century mathematics were finding their way into nearly all countries and cultures (see Chapter 2 of Volume 5). Today, people from all over the world take part in various mathematical conversations, and they tend to speak the same language (or the same "dialects," considering how difficult it is to communicate across sub-disciplines of contemporary mathematics). In the wake of decolonization, educators and historians have made a concerted effort to recover and revalue the respective traditions of their native cultures. In the case of mathematics, this has led to studies in the area of ethnomathematics, on the one hand, and to revisionist histories of mathematics, on the other. The latter emphasize, for example, the high level of algebraic achievement in Medieval China, or the sophisticated ideas on infinite series in the Kerala culture of India. These studies have also suggested the possibility that parts of this older knowledge may well have reached Europe and could have exerted a stronger influence on European mathematics than previously recognized, including the rise of calculus in the seventeenth century (see Volume 2).

A century ago, the realm of higher mathematics belonged almost exclusively to males. Even as educational opportunities for women gradually improved, chances to pursue a professional career remained very limited (see Chapter 2 of Volume 5). However, the emergence of "second-wave" feminism, inspired in part by Simone de Beauvoir's 1949 analysis in *The Second Sex*, led to calls in wealthier countries for full participation by females in all aspects of science and technology. Rooted in liberal ideologies, these voices not infrequently pointed to communist nations in which female participation was greater in some areas (notably medicine). The emergence of an influential feminist worldview (and, indeed, a more generally gender-inclusive one) has been gradual and ongoing.

Fueled in part by such thinking, the gradual establishment of various kinds of laws and rules against discrimination has gradually facilitated the entry of women into the Academy and other mathematical professions. Over the past five decades or more, the increasing participation of women has depended on a number of individual battles and victories, all amounting to the fact that women can function both as leaders and innovators in mathematics. Nonetheless, conservative worldviews—sometimes based on religion but often simply on sexism—continue to deny that such activity is appropriate for women. Such views are not absent from the world of research mathematics. In general, it has been very difficult for women to find their ways and become comfortable in the community of professional mathematicians. Great progress has been made in the last half-century, however, leading up to cases like those of Miriam Mirzakhani (Fields Medal) or Karen Uhlenbeck (Abel Prize), yet much more remains to be done.

Together with feminist criticism, growing concern with lingering postcolonial effects as well as ecological concerns (especially in relation to climate change) pose challenges to traditional notions about mathematical practice. These trends also reflect a renewed interest in philosophical and ethical issues as these relate to social and economic life.

If attempts to realize a monolithic understanding of mathematics have been abandoned, what are some of the newer directions that have since appeared? In

FIGURE 4.4. Stolperstein (stumbling stone) for Alexander Grothendieck, Brunnenstrasse 165, Berlin Mitte. Wikimedia Commons.

recent decades, mathematicians, physicists and computer scientists have taken part in an ongoing debate about the role of computers in the world of mathematics at large. A central question for many is whether there is a post-human future for mathematics. No one denies the relevance of computers and their profound impact on research in many areas, but there is a division of opinion about the long-term effects. In particular, some have prophesized the coming of a new era in which mathematicians present their findings very differently. They claim that the established style—a chain of theorems together with rigorously deductive proofs—will eventually die out, supplanted by computer-assisted proofs and probabilistic arguments. Other experts, while remaining within the game of rigorous proofs, nevertheless expect that

FIGURE 4.5. Front pages of Iranian newspapers bearing portraits of the Iranian-born mathematician Maryam Mirzakhani, who died of cancer on July 15, 2017, at the age of 40. In 2014, Mirzakhani became the first woman to win the coveted Fields Medal. Photo by ATTA KENARE / AFP / Getty Images.

computers will help formalize and verify whole areas of mathematics. Yet, some have a very different picture of what mathematics is and where it should go. Bill Thurston argued persuasively for the need to reinforce proof as understanding *for* and *by* human mathematicians (Thurston 1994).[20]

This ongoing debate among experts might appear far removed from the realm of ordinary worldviews and yet it clearly reflects an age-old problem. If mathematics represents the most secure form of knowledge human beings can possess, what forms does it take and what does it mean to have access to it? The notion of mathematics as superhuman, indeed divine, has a storied tradition (from Plato to Kepler, and from Gauss to Cantor). In today's secular culture, mathematics is wedded to science, not religion, and there are few who share what Einstein expressed when he spoke about his science as rooted in what he called "cosmic religion." Philosophers who struggle with the meaning of mathematical knowledge today are apt to think in terms of Alan Turing's universal computer, an idealized machine (see Chapter 6). One of the secularized incarnations of such conceptions is the idea of a fully objective proof-structure. "Fully objective" is understood here in an externalized sense: formal step-by-step proof, mechanically checkable, something like the perfect instrument for attaining mathematical knowledge for many technological optimists.

One can easily imagine computers capable of proving theorems that human beings will be unable to understand. In the view of Michael Harris, the basic unit of mathematics is the concept rather than the theorem, and the purpose of a proof is to illuminate a concept rather than merely confirm a theorem. For him, "the replacement of deductive proofs by probabilistic or mechanical proofs should be compared, not to the introduction of a new technology for producing shoes, say, but rather to the attempt to replace shoes by the sales receipts, or perhaps the cash profits, of the shoe factory" (Harris 2008: 972). Other authors have tried to elaborate a similar view, emphasizing that mathematics is about conceptual work. While acknowledging that some areas of basic mathematics enjoy certainty, they claim that in areas where certainty is absent a type of objectivity nevertheless prevails.

One need not be a technological pessimist to believe that mathematics should remain accessible to human beings. In fact, Thurston argued that the usual semi-formal, linguistic presentations of mathematical results (in the highly structured format "definition-proposition-proof") act as an *impediment* to understanding. He understood mathematics as a social activity, one in which informal communication played a vital part in imparting and mentally structuring the relevant material that experts master. This was particularly relevant in the more "visual" and "multidimensional" areas, geometry and topology, to which he contributed many important new insights. Much of the excitement his work generated came from his ability to communicate a richer *understanding* of mathematics.[21]

Although one can sympathize with Thurston's way of thinking, it seems likely that both pathways will develop and interact, since conceptual understanding and formal-symbolic expression play complementary roles in the game of mathematics, and probably both are needed. If so, then mathematics and mathematicians stand to gain much from renewed attempts to express key concepts and ideas in a deeper and richer way. Combining these with improvements in the symbolic presentation and formalization of results may then facilitate human–computer interaction and thus improve the checking and certification of results. This vision for the future is very much in keeping with a pluralistic conception of mathematical knowledge.

CHAPTER FIVE

Describing and Understanding the World

Quantum and Relativistic Physics

MARTA JORDI TALTAVULL AND TILMAN SAUER

The relationship between mathematics and physical theory in the twentieth century gained decisive shape through two major conceptual transformations. Quantum mechanics, which concerns the fundamental theory of matter, raised its head at the turn of the century and eventually consolidated in the years 1925–7. A decade earlier, the general theory of relativity emerged as a new theory of space, time and gravitation that revolutionized astrophysics and cosmology. Quantum concepts have since changed the way we understand the microscopic world. One of the key differences between classical physics and quantum theory is that the former assumed physical processes were always continuous and determined by universal natural laws, whereas the latter treats motions and energy exchanges at a microscopic level as discontinuous and fundamentally unpredictable. Relativity theory changed our understanding of the macroscopic world. Physicists came to realize how space and time intertwined and that spacetime itself shared in the dynamics of physical laws. Both transformations implied fundamental mathematical and philosophical changes in our way of describing and understanding physical reality.

Moreover, mathematics played a major role in both of these developments. Mathematical concepts and theories helped adapt physical theory to new experimental findings and empirical evidence as well as to react to problems that arose as inner-theoretical difficulties. The outcome, in each case, was a novel foundational equation, and to this day, these two fundamental equations govern our understanding of the physical world, both small and large. The Dirac equation or its non-relativistic counterpart, the Schrödinger equation, represents the fundamental law for all of quantum physics. This includes atomic and molecular physics, nuclear as well as elementary particle physics. The Einstein equation serves as the foundation for the general theory of relativity and, thus, for modern gravitational theory. It shapes our understanding of the cosmos in the large as well as its genesis and evolution.

Aside from their fundamental importance, these equations have a larger significance that points to a comparatively recent understanding of how mathematical ideas relate to physical theories. First, as noted above, both of these equations captured major transformations in the basic physical concepts underlying our understanding of the world. Second, both also involved the introduction of new mathematical techniques and formalisms into physical theories. Third, in both cases it took many years and much effort to elaborate the consequences and implications inherent in those equations and to understand fully their import for our present worldview. In this process, new mathematical concepts and a new physical semantics evolved together to reflect the new ways of describing and understanding physical reality. In the quantum case, the most prominent new concepts were the notion of a Hilbert space and its corresponding wave functions. In the case of relativity theory, the notions were metric tensors, affine connections and the associated concept of spacetime curvature.

The advent of these fundamental equations thus came with a respective physico-mathematical framework. Not only did scientists take up and adopt various novel mathematical ideas, they remained faced with the challenge of exploring the consequences hidden within the compact notation of a single equation. Major developments in twentieth-century mathematics and physics resulted from efforts aimed at understanding those implicit consequences and clarifying the physical meaning of the mathematical concepts. Mathematical physicists introduced new concepts, reworked existing ones, investigated the space of solutions, enhanced their predictive power, clarified the connections with experiment, reflected on the unity and consistency of theories, extended them to other physical contexts, and set the grounds for new ways to model and simulate nature. These aspects reflect how the mathematical sciences of the twentieth century involved multifaceted, exciting activities that included modeling and problem solving, while at times making unexpected discoveries.

Perhaps in the twentieth century more than ever before, the availability of all-encompassing fundamental equations transformed physical theorizing to a goal-oriented enterprise of modeling nature along lines guided by solutions of those equations. Models of the nucleus and larger structure of atoms in the microscopic realm as well as models of the cosmos at large and its development gave meaning to the equations. They provided these mathematical equations with physical semantics and thereby mediated their realist intent. In turn, the fundamental equations determined key properties of the models by providing particular solution spaces, which guided the elaboration of the models.

The ubiquity of models and modeling in modern physical theory is by no means restricted to quantum and relativistic theories. Mathematics had a pervasive influence in many other fields of physical theory during the twentieth century, including the development of statistical mechanics and ergodic theory. It continues to play a major role in fluid mechanics, based on models for solutions of the Navier-Stokes equation. One can point to all kinds of simulation efforts, not only in materials science but also in chemistry and biology, such as neuronal networks, population dynamics, and other complex systems. All these attest to the significance of modern mathematics for present-day scientific understanding of the world in which we live. In order to capture these complex and multifarious processes, this chapter will focus on the interaction between mathematical developments and physical theory by concentrating on theory development in quantum physics and relativity theory. For reasons of space, we will not take other branches of the physical sciences into consideration and only scarcely discuss experiments.

The chapter consists of seven sections organized as follows. In the first, we describe the emergence of the special theory of relativity and the development of the old quantum theory up to 1916 as preconditions for the major transformations that would build on them. The second section explains the emergence of quantum mechanics in 1925–6 as a consistent theory of the microscopic world, highlighting its relationship with new mathematical concepts and theories and taking the Schrödinger equation as a central turning point. The third section then deals with the interpretative and philosophical problems associated with the new quantum mechanics. The mathematical and conceptual problems derived from attempts to solve Schrödinger's equation for various atoms and molecules forms the central theme in the fourth section, whereas the fifth outlines further developments in which the ubiquitous relevance of quantum mechanics for all fields of physics became increasingly apparent. In the sixth section, we discuss transformations in our understanding of the world at large based on the general theory of relativity after Einstein first launched it in 1915. Many of its consequences and implications only became evident decades later through efforts to elaborate on solutions of the Einstein equations. In addition to the three classical tests of general relativity, we also consider the recent successful confirmation of the Einstein equations by the direct detection of gravitational waves. We conclude in the seventh section with a brief discussion of the role of both special and general relativity in cosmological applications as well as for attempts to unify fundamental physical theories.

SPECIAL RELATIVITY AND EARLY QUANTUM THEORY: 1900–16

The quantum hypothesis made its first appearance at the very beginning of the twentieth century, when the German theoretical physicist Max Planck introduced it in 1900. Planck went on to become a doyen among German scientists, but back at that time, he was a young professor in the Prussian capital of Berlin. In fact, it was not until 1925–6 that the new quantum mechanics made a complete and radical departure from classical mechanics, as we will describe later. From 1900 until the early 1920s, theorists developed microphysics by amalgamating the new quantum hypotheses with models and mathematical tools borrowed from classical physics. They later referred to the work from this period as the so-called "old quantum theory."

After Planck, Albert Einstein further elaborated the quantum hypothesis in 1905, while employed as an unknown patent clerk in the Swiss capital of Bern. Both Planck and Einstein worked in the field of mathematical or theoretical physics, a relatively new sub-discipline in the physical sciences pursued by theoreticians (Jungnickel and McCormmach 1986). These researchers focused on the theoretical and mathematical understanding of physical theories rather than carrying out experimental investigations that aimed to test or elaborate on them.

Planck was dealing with an old physical problem that had arisen at the borderline between the theory of heat (i.e., thermodynamics), and electricity and magnetism (electrodynamics). He was trying to understand the phenomenon called black body radiation. A black body is an idealized non-reflecting body, which absorbs all the light and radiation it receives, while emitting electromagnetic radiation in such a way that it remains in thermodynamic equilibrium with the electromagnetic radiation field. What made black body radiation particularly intriguing for theoreticians was the fact that this behavior is associated with a universal function that at any given temperature relates the radiation intensity to its frequency.

Planck introduced his quantum hypothesis in order to derive a new—and as it turned out correct—radiation law, which perfectly matched earlier experimental findings obtained by experimental physicists in Berlin. He found this new expression for the radiation law by means of a theoretical derivation based on Ludwig Boltzmann's statistical methods in thermodynamics. This approach related macroscopic properties of matter to microscopic configurations by means of probabilistic considerations. Planck followed Boltzmann's calculations closely, except for one crucial detail: Boltzmann allowed for molecules to have any value of energy. Instead, Planck conceived of microscopic resonators whose energy was a sum of quanta of energy that were not arbitrarily small. Their size was finite and proportional to frequency v via universal constant h, now known as Planck's constant. Distributing these quanta of energy over microscopic resonators was the crucial step that allowed Planck to derive his new radiation law.

Despite the far-reaching consequences that later arose from Planck's introduction of the constant *h* and discrete quanta of energy, their deeper implications were only unveiled much later. Planck had no intention back in 1900 of revolutionizing physics; much to the contrary, he conceived of quanta of energy merely as a tool for calculation. Quantum discontinuity was thus born almost accidentally, as it were, and physicists viewed this as a curious anomaly rather than a threat to their fundamental theories. It took some time before they comprehended the enormous consequences Planck's idea would pose.

Einstein was among the first to realize that Planck's energy quantization implied a radical break with classical physics.[1] Physicists had always imagined the absorption and emission of energy as continuous, never as a process involving quantization. In 1905, Einstein, while attempting to understand Planck's law of black body radiation, proposed that light, or electromagnetic radiation in general, behaved as though it consisted of discrete light quanta of energy $h\nu$. Most physicists long treated this bold hypothesis with reserve, however. Even in 1922, when Einstein received the Nobel Prize, the Nobel committee mentioned neither relativity nor his theory of light quanta in its award citation (instead, they cited his work on the photoelectric effect). Nevertheless, Einstein's contribution served to fortify the conviction that one could not discount the importance of quantum discontinuities in physical phenomena.[2]

The history of relativity theory began in that same year of 1905 when Einstein published, within a few weeks after submitting his paper on the light quantum hypothesis, another article that would have an immense impact on physics. Entitled, "On the Electrodynamics of Moving Bodies," it gave the first exposition of what Einstein later called the special theory of relativity. Because of these two remarkable papers, in addition to a few others published that same year, historians have dubbed 1905 Einstein's miracle year in physics. The term calls to mind Newton's discoveries during the Plague of 1665–6, which he escaped by returning to his home village of Woolsthorpe, where he famously arrived at his first insights into the universal law of gravitation, the spectral composition of light, and the basic ideas for his version of the differential and integral calculus. Einstein, too, was a young and still rather unknown theoretical physicist, who spent his weekdays working in the Swiss Office of Intellectual Property. In his 1905 paper, he solved a long-standing puzzle that first arose toward the end of the nineteenth century.

For many decades after Newton published his *Principia* in 1687, his theory of mechanics dominated physical theorizing. In it, he posited an absolute space filled with material particles, which moved about under the influence of inertia and mutually attractive forces. Newtonian classical mechanics served as the paradigmatic physical theory for some two hundred years, but by the end of the nineteenth century another fundamental theory arose that seemed to complement this world picture. Following the theory of electrodynamics introduced by James Clerk Maxwell, physicists conceptualized electric and magnetic forces as space-filling fields. Charged particles created these fields and they themselves were subject to the moving forces of those fields, which depended on their momentary position.

An important feature of Maxwell's electrodynamics was the assumption of an all-pervasive ether, which acted as the medium for the propagation of electric and magnetic waves. In Maxwellian electrodynamics, no physical phenomena could propagate faster than the speed of light, and the assumption of an ether seemed to imply the existence of a preferred reference system, i.e., the one in which the ether was at rest. Indeed, in the influential electrodynamic theory of the Dutch physicist Hendrik A. Lorentz, the notion of an absolute ether came to serve as a substitute for Newton's long-disputed concept of absolute space. Nevertheless, the ether hypothesis stood in conflict with another basic assumption: the principle of relativity in classical mechanics. This states that all frames of references in constant relative motion are essentially equivalent, since each such inertial frame leads to the same form for Newton's laws of motion. Yet, this principle would seem to imply that there could be no one distinguished absolute frame of reference.

Einstein solved this apparent contradiction, in effect, by modifying the principle of relativity in mechanics and dispensing with the notion of an absolute ether. His solution was at once elegant and radical, in that it involved a reflection on the epistemological status of the notions of space and time and, in particular, the notion of distant simultaneity. Einstein's key insight fastened on realizing the implications of the fact that the speed of light c is a universal constant. This meant that one needed a new law for the addition of velocities of two reference frames in uniform motion. Realizing that the assertion of simultaneity in two reference frames in uniform relative motion crucially depends on the finite speed of light, Einstein showed by dropping the notion of an absolute ether frame, one could still take the Lorentz coordinate transformations as the basic kinematic feature of spacetime.

While most physicists soon accepted the special theory of relativity as a logically consistent, empirically adequate theoretical framework for phenomena involving high velocities, it took much longer for the light quantum hypothesis to gain acceptance within the physics community. During the ensuing years, however, Planck's new constant popped up in various unexpected places, after which physicists began to take the quantum hypothesis seriously. In 1907, Einstein extended quantum discontinuity to account for the low-temperature behavior of the specific heats of solids, by modeling a solid as a set of harmonic oscillators with energies quantized like Planck's resonators. Later, in 1911, the German pioneer of physical chemistry Walther Nernst applied quantization to gases, again with satisfactory results in comparison with experiments. The growing interest in quantum physics was apparent in November 1911, when the first Solvay Conference took place in Brussels; it was devoted to the general topic of "Radiation and the Quanta." This was the first in a series of highly significant international conferences on theoretical physics funded by Ernest Solvay, a wealthy Belgian industrialist and philanthropist who was interested in the new theoretical physics of his time.

Another turning point in the early history of quantum physics was the publication in 1913 of the first quantum model of the atom by the young Danish theoretical physicist Niels Bohr. Up to this time, physicists had proposed various atomic models, but none of these incorporated the quantum hypothesis. In the context of classical physics, Joseph J. Thomson had proposed a model with negatively charged electrons embedded in a positively charged mass. Afterward, Ernest Rutherford conducted some famous experiments, which showed that the atom's mass was concentrated in a very small, positively charged nucleus. This led him to propose a new planetary atomic model. In analogy with our solar system, its nucleus formed the center around which orbited the much smaller negatively charged electrons. Although this model accorded well with certain radioactive experiments, it had one very serious drawback: from the point of view of classical physics, it was mechanically unstable. In electrodynamics, electrons continuously radiate energy as they move. Thus, in Rutherford's model they would lose energy and eventually fall into the nucleus.

Niels Bohr borrowed from Rutherford's atomic picture, but at the same time, he introduced a radical break with classical physics. Bohr avoided the problem of instability by simply postulating that electrons remain in stable states on their orbits around the nucleus. Furthermore, he also postulated that not all possible orbits were valid, only those with certain discrete energies E_n, characterized by an integer number n. Here again, a quantum discontinuity entered the picture, and n later was called the first quantum number of atomic physics.

Bohr also came up with a second postulate, which was just as impossible from the point of view of classical physics. He stipulated that electrons would absorb or emit energy only by means of sudden quantum jumps between their discrete stationary orbits. Hence, the frequency of the emitted or absorbed radiation depended on the energy difference between the initial and final states, in accordance with the relation $E_1 - E_2 = hv$, in which E_1 and E_2 are the energies of the stationary states and v the radiation frequency. Once again and rather surprisingly, Planck's constant h took on a new meaning. According to classical physics, light frequency related to the frequency of revolution of the orbiting electrons. In Bohr's atomic model, on the other hand, the frequency of light depended on a process—quantum jumps between stationary states—that clashed with the principles of classical physics. Bohr's second postulate simply cast this difficulty aside in order to account for the absorption and emission of light by atoms at specific discrete frequencies, which he related to the material elements under study, as characterized by their spectra.

Bohr supported these postulates with impressive experimental evidence. He was able to reproduce numerically with a high degree of accuracy the observed frequencies of the spectral lines of the monoatomic gas of hydrogen (its Balmer series). This was an extraordinary feat, which depended in an essential way on his two postulates, and it left many of his colleagues flabbergasted. Nevertheless, during these early years, Bohr's theory had many detractors. Especially his second postulate appeared to stand in conflict with classical notions of causality. No one could provide a mechanism to explain or describe the quantum jumps. It was as if one had to ascribe an apparent "free will" to the electrons in order to account for their jumping between stationary states.

This situation changed, however, after 1916 with further advancements on Bohr's atomic theory achieved by the prominent German theoretical physicist Arnold Sommerfeld, director of the Institute of Theoretical Physics in Munich. Drawing on resources in classical mechanics, in particular celestial mechanics, Sommerfeld placed Bohr's atomic theory on a strong mathematical footing. He generalized Bohr's circular orbits to ellipses and introduced a second quantum number to accommodate various degrees of freedom. Sommerfeld's new mathematical formulation allowed an extension of Bohr's theory to other spectral phenomena. He was also able to explain the fine structure of the observed line spectrum of hydrogen. Realizing that their velocities around the nucleus of the atom were near the speed of light, Sommerfeld treated the motion of the electrons on quantum orbits relativistically. This breakthrough achievement represented the first application of Einstein's special theory of relativity to the new quantum theory based on hypotheses first introduced by Planck and afterward extended by Bohr.

Following these early successes, quantum theory gained many adepts, even though it remained far from complete. By the late 1910s, it was still mainly an amalgam of methods from classical mechanics mixed with quantum rules. Furthermore, no one could explain a number of phenomena closely related to quantum theory. For example, physicists could predict the location of the lines in the frequency spectrum, but they could not account for why some were stronger, whereas others appeared to be rather weak. An explanation of the intensities of the spectral lines thus eluded the early quantum theorists. There were also continuous phenomena, already successfully described in classical terms, but about which quantum theory had nothing to say. An example was optical dispersion, the continuous spread of light into different colors when passing through certain materials.

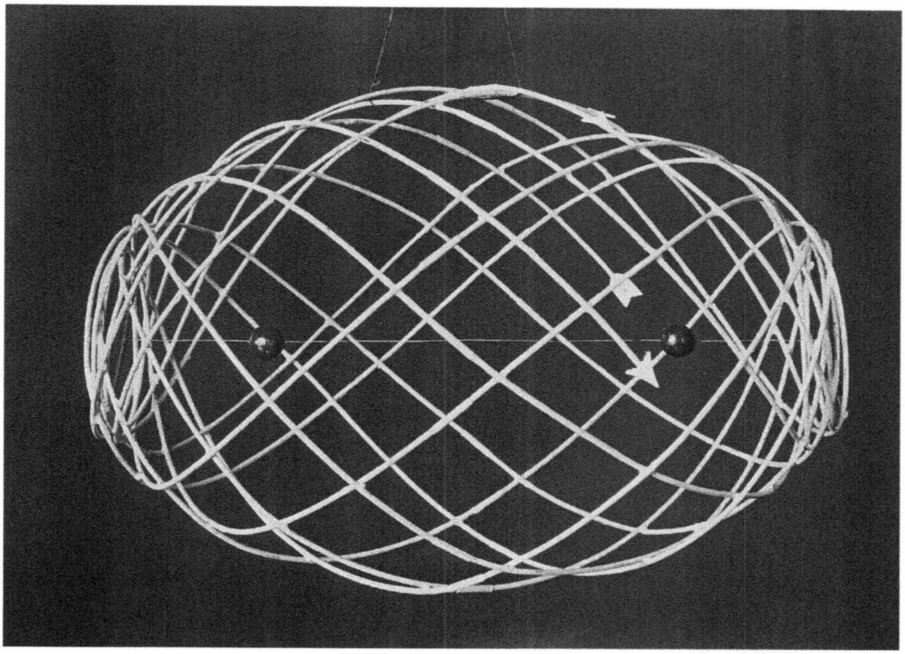

FIGURE 5.1. Model of a hydrogen ion. Photo: Deutsches Museum, Archiv, BN 02935.

In 1918, Niels Bohr put forward his correspondence principle in an attempt to construct an all-encompassing account of the emerging quantum world. He based his principle on the idea that classical and quantum physics should coincide for large quantum numbers (large values of n). Bohr proposed using classical tools to represent electron trajectories initially. In particular, the electron's motion could be represented as a Fourier series of harmonics (see Chapter 3 of Volume 5), much like a classical resonator whose motion was a sum of the contributions of all its harmonics. Here, one could utilize tools from analytical mechanics, such as Hamilton-Jacobi methods and differential equations, while treating interactions with light as mechanical perturbations with electromagnetic radiation. Only toward the end would the theorist pass from classical to quantum physics by substituting classical frequencies for quantum frequencies (transitions) and replacing the squares of harmonic amplitudes by the probabilities for the corresponding quantum transitions. In the limit of high quantum numbers, the two versions should then coincide. To be sure, Bohr's methodology was controversial and not without competitors, but it played an important role in the developments of the quantum revolution of 1925–6.

FIGURE 5.2. Niels Bohr and Albert Einstein, Leiden 1925. Photo by Photo 12 / Universal Images Group / Getty Images.

EMERGENCE OF QUANTUM MECHANICS

In the years 1925–6, the patchwork of classical and quantum hypotheses—the old quantum theory as it soon came to be known—was replaced by a new quantum mechanics. This was the period of the quantum revolution, properly speaking. During this time, two versions of quantum mechanics arose nearly simultaneously and largely independently from each other: matrix mechanics and wave mechanics.

Bohr's correspondence principle determined the route that eventually led to matrix mechanics, though the German physicist Werner Heisenberg took the first step in this direction. Heisenberg belonged to a younger generation of ambitious and mathematically inclined theoreticians who, while building on the foundations of the old quantum theory, transformed it into the new quantum mechanics. At the time, he was working at Bohr's Physics Institute in Copenhagen, before joining Max Born at the University of Göttingen. In keeping with the correspondence principle, Heisenberg's point of departure was the Fourier series representing the motion and momentum of electrons in classical theory. He assigned to each term of the Fourier series (harmonic frequency and amplitude) a quantum meaning in terms of transitions and their probabilities. More specifically, he gave each harmonic frequency and amplitude two indexes, one referring to the initial state of the quantum transition, and the second to its final state. This seemingly simple innovation had far-reaching consequences, in particular: it turned out that when one multiplied two such physical quantities, the answer depended on their ordering, i.e., $AB \neq BA$, so multiplication in this setting was no longer commutative. One year later, Heisenberg's colleagues in Göttingen, Max Born and Pascual Jordan, realized that these physical quantities obeyed the same multiplication rules as matrices. This meant that physical observables were not simple scalar quantities, i.e., numbers for which the commutative law of multiplication held; instead, matrices were suitable for their mathematical representation.

At nearly the same time, in 1926, the Austrian physicist Erwin Schrödinger introduced a second version of quantum mechanics. Schrödinger was then working in Zürich, but he later succeeded Planck at the University of Berlin. He was among those who expressed discomfort with the unintuitive nature of matrix mechanics, particularly its lack of visualizability, and so he sought an alternative approach. Instead of drawing on Bohr's correspondence principle, Schrödinger took his inspiration from the French physicist Louis de Broglie, who had proposed the idea that just as light sometimes behaves like a particle, so can matter exhibit properties that resemble a wave. Schrödinger therefore set out to develop a wave theory of matter, and he eventually obtained a differential equation that described the temporal evoluton of such hypothetical matter waves Ψ:

$$i\hbar \frac{\partial \Psi}{\partial t} = H\Psi.$$

The solutions of this equation took the form of a series of proper wave functions associated to proper energy values associated with a Hamilton operator H. Each proper energy value corresponded with the energy level of one stationary state (or several in certain cases of degeneracy) and it was associated, in turn, with a wave function.

Schrödinger's equation received a more enthusiastic response than the one accorded Heisenberg's matrix mechanics, no doubt due to its more intuitive appeal. His differential equation looked appealing to those familiar with classical mechanics, since it suggested a more intuitive space-time description of microscopic events than that offered by abstract matrix mechanics. Nevertheless, as Schrödinger himself later demonstrated, both formalisms led to the same experimental predictions, in particular for the spectrum of the hydrogen atom.

Why do these two different approaches lead to the same results? It took a few years to answer that question, but the equivalence of the two formulations became clear once both were seen in the light of a third mathematical formulation. As it turned out, matrix and wave mechanics share an algebraic structure represented at its core by a special kind of vector space, even though neither theory explicitly presupposed or relied on the notion of vector spaces. It was while exploring the consequences of the wave equation that several physicists—in particular, Fritz London, Pascual Jordan, and Paul Dirac—began to probe the underlying mathematical structure of Schrödinger's wave mechanics. Then, in 1927, John von Neumann, in a series of three papers, definitively proposed using these new mathematical spaces as a framework for quantum mechanics. More than a decade earlier, the prominent Göttingen mathematician David Hilbert had already developed much of this theory, today called Hilbert space theory. Hilbert spaces are similar to Euclidean spaces, but have infinitely many dimensions. Following von Neumann, quantum theorists began to think of wave functions and stationary states as represented by vectors in a Hilbert space. Physical quantities were then operators, represented by matrices, that acted on the vectors in a Hilbert space, and thus on the stationary states. This was an entirely new mathematical representation of fundamental concepts in the theory of matter, and thereafter physicists had to familiarize themselves with this new mathematics in order to understand the foundational structure of quantum mechanics.

PROBLEMS OF INTERPRETATION OF QUANTUM MECHANICS

If physicists felt challenged by the mathematical underpinnings of the new quantum mechanics, its physical interpretation also proved especially difficult. In 1926, Born proposed an interpretation commonly accepted today: he interpreted the square of the wave function as the probability of finding an electron in a certain position or with a definite momentum. This viewpoint constituted a fundamental break with respect to the application of statistical methods in classical physics, which did not preclude the possibility of determining both the momentum and position of each particle with arbitrary precision. Quantum mechanics, on the other hand, excluded the possibility of obtaining such exact information as a matter of principle. All that one can

FIGURE 5.3. The Fifth Solvay Physics Conference in Brussels, October 1927. Delegates included Marie Curie, Paul Dirac, Albert Einstein, Paul Langevin, Max Planck, and Erwin Schrödinger. Photo by SSPL / Getty Images.

know about a quantum system is the probability of finding a certain state if we perform a measurement, and this posed a fundamental challenge to the classical notion of causality.[3]

The gap between classical and quantum mechanics was set into stone by Heisenberg's uncertainty principle, published in 1927. This principle follows from the non-commutative algebra underlying quantum mechanics in connection with two conjugate quantities, such as a particle's position and its momentum. Heisenberg interpreted this non-commutativity as imposing a limit on the precision with which the values of the two conjugate variables could be experimentally determined at the same time.

Bohr went further still. Nature, for him, expressed herself in complementary ways, exhibiting momentum and position as conjugate quantities. This principle of complementarity also applied to other properties of nature, such as the corpuscular and wave descriptions. Light was thus both a discrete corpuscle and a continuous wave. An experiment designed to measure the wave properties of light, such as diffraction, would exclude measuring its corpuscular properties, such as the photoelectric effect, and vice versa. Thus, according to Bohr, it was impossible to isolate the outcome of an experiment from the very action of the observer who sets up a specific kind of experiment by choosing a specific quantity to measure. The observer thus becomes an inextricable part of the experimental apparatus itself.

This new understanding of the very nature of experiment presents to this day a major difficulty for philosophers of physics, who refer to this as the measurement problem in quantum mechanics. The mathematical structure of Schrödinger's wave equation implies that a superposition of solutions is also a solution. This implies that the temporal evolution of the wave function represents a continuously determined process. Following a measurement, just one possible solution suddenly manifests itself, an irreversible process that discontinuously destroys any prior superposition. It thus leaves the system in the exact state corresponding to the measured value of the observable. Before determining its precise position, the electron is extended over space as represented by a wave function. Once she/he actually measures its position, this wave function reduces to the single solution representing the position found.

This process—known as wave function collapse—has no counterpart in classical physics. Schrödinger's famous thought experiment with a cat, proposed in 1935, epitomizes this feature. We imagine a cat sitting in a closed box and a macroscopic mechanism that releases a toxic substance, which may leave the cat either dead or alive inside the box. A random event, such as the decay of a radioactive atom, will trigger this mechanism. Prior to performing a measurement, the radioactive atom is in a superposition of two possible states, decayed and not, which means that the cat, too, is in a superposition of two different states, dead and alive. Only after performing the measurement, namely, opening the box and looking inside, will the wave function collapse and determine one of these two states.

The superposition principle has a close connection with another quantum phenomenon: entanglement. Two or more particles, even though very far away from one another, may nevertheless be in an entangled state of mutual dependence. When this occurs, a measurement of the state of the first particle results in a function collapse that unavoidably affects the final state of the second. These theoretical features constitute core assumptions of the Copenhagen interpretation of quantum mechanics, so named because Niels Bohr and members of his school propagated them. Despite its inherent difficulties, the Copenhagen interpretation is still widely accepted today, though some rather popular alternative interpretations of quantum mechanics offer very different answers to the problems just mentioned. These include the many-worlds interpretation, quantum decoherence, and the hidden-variables theory. However, discussion of these alternative interpretations of quantum mechanics falls beyond the scope of this chapter.

Even during Bohr's lifetime, not all leading physicists agreed with him, the most famous heretic being Albert Einstein. Although Einstein had pioneered early quantum theory, he gradually distanced himself from fully accepting its implications, once he realized the profundity of the break between quantum and classical physics. Ultimately, he rejected the standard way of understanding the world through quantum physics because he could not accept a description of physical reality that was fundamentally probabilistic. He exemplified this attitude with his famous saying: “God does not play dice.” Entanglement was another aspect of quantum theory that Einstein found intolerable. He pointedly laid out the difficulty in a famous paper, which he published together with Boris Podolsky and Nathan Rosen in 1935. The crux of the matter, soon called the EPR-paradox after the initials of the three authors, was how to reconcile the notions of causality, reality and separability in physical systems, a topic of much debate in the philosophy of physics ever since. Contra Einstein, in recent years, entanglement has become a constitutive feature of the modern understanding of quantum mechanics. Today, physicists try to exploit its characteristic features in quantum cryptography, quantum computation, and in other potential applications of modern quantum physics.

SOLVING THE SCHRÖDINGER EQUATION

Initially, one could obtain exact solutions of Schrödinger’s equation only for the simplest systems, namely those with a single electron. This meant, in effect, studying a single element in the periodic system, namely the hydrogen atom, which gave rise to the already-mentioned Balmer series. Thus, in order to study even slightly more complicated systems, one needed to employ approximation methods.

Knowing this, Max Born and the American physicist J. Robert Oppenheimer set out in 1927 to calculate the structure of molecular hydrogen, i.e., a system with two protons as well as two electrons. Their approximation scheme set the grounds for the quantum mechanical treatment of molecules for decades to come. In particular, Born and Oppenheimer assumed one could treat the respective contributions from the nuclei and the electrons separately, owing to the large difference between their masses. This assumption was implicit in the method introduced in 1928 by the English mathematician and physicist Douglas R. Hartree for calculating molecular orbits. Hartree’s basic idea was to obtain a good approximate solution through iteration, a method still in use today, although in a modified and improved form. This scheme brought to the fore a sharp break with Bohr’s atomic model in the old quantum theory. Whereas Bohr had still assumed that electrons would move around the nucleus on clearly defined orbits, the picture of matter implied by the Schrödinger equation suggested cloud-like orbitals, regions in the atom where the probability distribution for the position of the electron was high.

Hartree’s method proved to be influential; when the Russian physicist Vladimir A. Fock later improved on it, he included the concept of spin in his

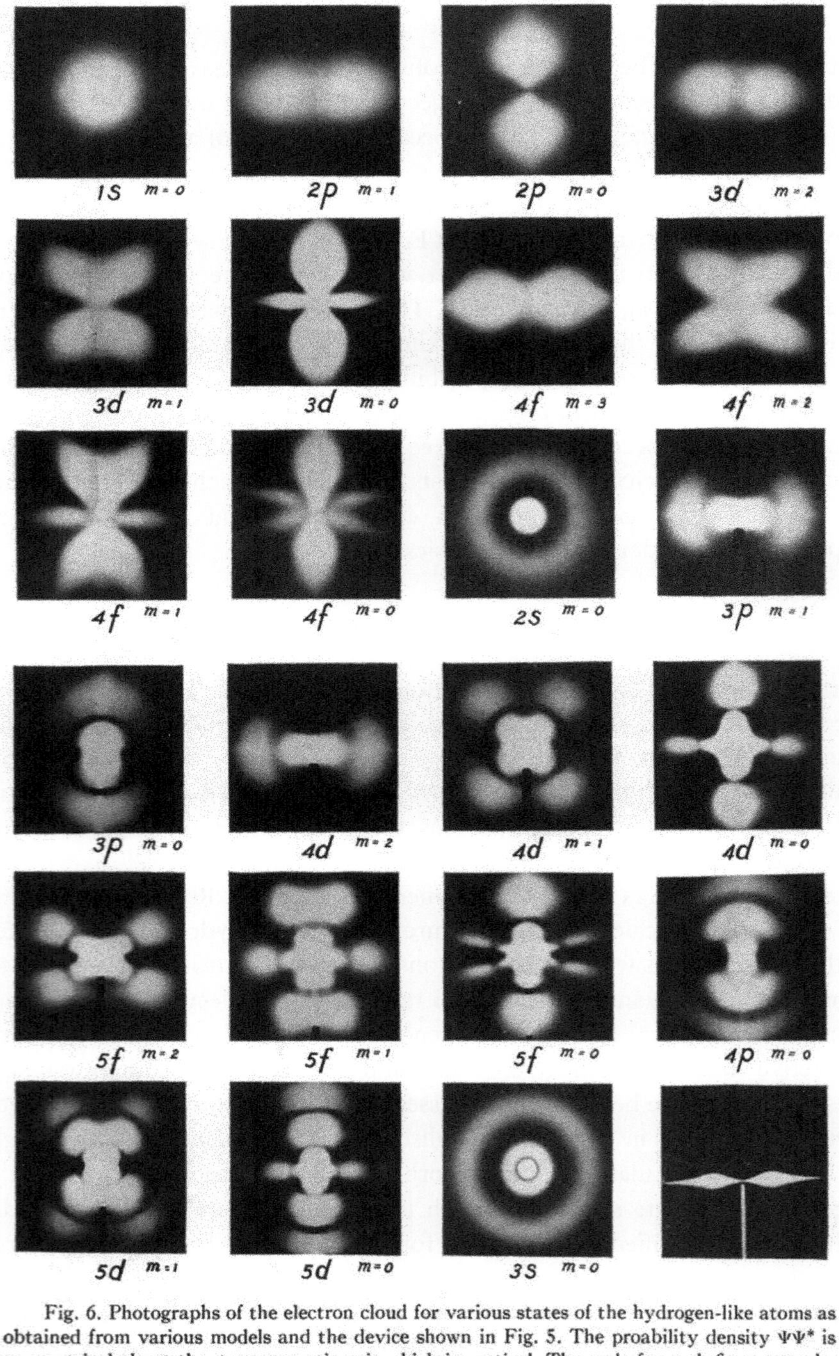

Fig. 6. Photographs of the electron cloud for various states of the hydrogen-like atoms as obtained from various models and the device shown in Fig. 5. The proability density $\Psi\Psi^*$ is symmetrical about the ϕ or magnetic axis which is vertical. The scale for each figure may be obtained from Fig. 4.

FIGURE 5.4. "Early graphical illustration of atomic orbitals." From: H. E. White, "Pictorial Representations of the Electron Cloud for Hydrogen-Like Atoms," *Physical Review*, 37 (1931), pp. 1417–25, Fig. 6.

calculations. Spin was another genuinely quantum mechanical concept which had emerged in 1925–6. Experimental results had suggested that the number of electrons in each atomic shell was limited. In 1925, the Austrian physicist Wolfgang Pauli proposed an explanation by means of his exclusion principle: Pauli assumed that electrons have an internal degree of freedom given by their spin, a quantum number, which could take on two discrete values. He further postulated that electrons only co-exist in an atom if they differed in at least one of their quantum numbers. Since two electrons with the same energy and orbit share the same quantum numbers, they must differ in spin. The name reflects the origin of the concept, although nowadays one conceives of electrons as point-like and thus they do not actually spin like extended bodies.

Pauli's exclusion principle had far-reaching consequences for the theory of matter. The fact that in any system two electrons with the same quantum numbers and the same spin could not occupy the same quantum state meant that the function describing the electron system had to be an antisymmetric solution of Schrödinger's equation. Systems in which the exclusion principle no longer held corresponded to symmetric solutions of Schrödinger's equation. Along with the notion of spin, another new mathematical framework emerged in the late 1920s: the mathematical theory of groups. Using group theory, physicists could analyze solutions of Schrödingers' equation to determine various combinations, symmetric and antisymmetric, of individual wave functions. Such combinations led to descriptions of different kinds of multi-electronic systems for more than two electrons.

The Hungarian theoreticians Eugene Wigner and John von Neumann, both then in Göttingen, took the decisive step of introducing group theory in quantum mechanics in 1926 and 1927. Wigner also promoted representation theory to study the ways the elements of a group operation act on a vector space. The theory of group representations bridges the abstract mathematical theory of groups with their actions on specific vector spaces, such as Hilbert spaces. In the 1950s, a revamping of the mathematical theory of groups in quantum physics took place in the context of elementary particle physics. Since that time, groups, symmetries and invariances have become standard topics in physics.[4]

EMERGENCE OF QUANTUM FIELD THEORY

In 1927, not only matter, but also light became an object of new quantization schemes. In that year, Paul Dirac put forward a theory of light-matter interactions, in which he quantized both matter and radiation. Dirac regarded the free classical electromagnetic field as a superposition of modes of harmonic vibration, which arises as the classical limit of quantum electromagnetic fields. The trick he introduced involved replacing each classical mode by a quantized harmonic oscillator of the same frequency. In order to mathematize the process, Dirac defined two new operators that acted on the quantum states of the quantized oscillators. These were the creation and annihilation operators, which either increased or lowered the number *n* of photons (light corpuscles) in each quantum state.

One year later, Dirac made another epoch-making contribution by introducing a relativistic equation, in which the creation and annihilation of matter emerged as an intrinsic feature of the formalism. Dirac's equation resulted from his attempt to bring Schrödinger's equation into harmony with the principles of special relativity. Schrödinger's equation, successful though it was, stood apart from the fundamental assumptions of modern spacetime physics, since it treated the time coordinate differently than the spatial coordinates. In order to remedy this difficulty, Dirac generalized the Schrödinger equation to produce the Dirac equation. Although its very formulation required advanced mathematical concepts, like matrix representation theory, Dirac's fellow physicists immediately celebrated his discovery. Using Dirac's equation, they could directly derive the concept of spin; moreover, the equation also explained hitherto puzzling experimental results pertaining to the fine structure of the hydrogen spectrum.

Yet, it also brought forth a serious difficulty: its mathematical structure implied the existence of states of negative energy, though it was unclear whether these corresponded to a possible state of matter. Dirac proposed that the negative-energy states were actually vacancies (holes) in a sea of positive energy states. Yet, these holes seemed to behave like particles; indeed, they manifested themselves as a new kind of particle with the same mass as electrons but with a positive charge. Eventually, they gained acceptance as positrons, a new kind of elementary particle, joining electrons and protons as the only two then known fundamental particles. Moreover, the positron was the first particle whose existence arose solely on theoretical grounds, since it emerged from its mathematical representation rather than an experimental test.

Dirac thought that the creation of positrons resulted from the collision of gamma rays, each pair of which gave rise to one electron and one positron, an idea that eventually led to the postulation of other "antiparticles." In 1932, the American physicist Carl Anderson conducted experiments with cosmic rays (high-energy rays), during which he detected particles that were later identified as positrons. Around this time, physicists began to accept that particle creation and annihilation were novel features of Dirac's synthesis of quantum theory with special relativity theory. This gave rise to the postulation and experimental detection of other particles, such as neutrinos.

Quantum field theories, including Dirac's quantum electrodynamics, suffered, however, from an imposing theoretical difficulty. When studying high-energy interactions, the formalism inevitably led to infinities that were both difficult to understand and seemingly impossible to handle mathematically. It was not until the late 1940s that theoretical physicists found a way to resolve this problem by means of the so-called renormalization theory. The success of these efforts helped restore physicists' faith in quantum field theories, which many viewed with skepticism up until the 1970s. Three protagonists were mainly responsible for the emergence of renormalization theory in the late 1940s: the Japanese physicist Shinichiro Tomonaga and two American colleagues, Julian Schwinger and Richard Feynman. In their hands, renormalization developed into a collection of mathematical techniques used to compensate for infinities by introducing a cut-off parameter in the calculations (Schweber 1994).

In the 1970s, these ad hoc mathematical methods took on a new conceptualization when renormalization procedures were interpreted realistically. According to this new view, each energy scale corresponded to one single field theory that affected only those specific particles and types of interactions that were relevant in the given scale. This approach divided physical reality, as it were, into different quasi-autonomous domains, each governed by a different effective field theory. Some physicists thought that different effective field theories were definitely required for different scales, whereas others expected that a fundamental unifying theory for renormalizable high-energy fields should emerge that incorporated the established scaled theories. In any case, renormalization theory helped to make quantum field theories manageable and to legitimize this approach for understanding the behavior of matter and radiation from the standpoint of special relativity and quantum theory. These long-term developments eventually led to the current consensus understanding of photons and all elementary particles (protons, neutrons, electrons, muons and neutrinos) as localized excitations of different underlying fields.

Particle physics also gained support during the latter half of the twentieth century from growing recognition of the importance of mathematical theories associated with the concept of symmetry. This notion played a prominent role in the discourses on the construction of fundamental theories in physics from the 1950s onward, as awareness of the connection between symmetries and conserved quantities spread. There had been an earlier recognition of this in the classical mechanics of the nineteenth century, when mathematicians employed variational principles to derive conservation laws, such as conservation of energy or momentum. In purely mathematical language, however, the more general concept of invariance proved particularly relevant for modern field physics. In the 1870s, Felix Klein highlighted the importance of invariant quantities for the classification of geometries by means of their transformation groups. Around the same time, Sophus Lie undertook the first detailed study of continuous transformation groups and their associated invariants. These developments took place, however, with few direct connections to physical theories, in particular field physics.

The more general understanding of the relationship between conserved quantities and the existence of symmetries first emerged in early work on Einstein's general theory of relativity, as described in the section that follows. In that context, Emmy Noether proved two fundamental theorems in 1918, the first of which came to play a major role in modern particle physics. This states that in physical systems, represented by a variational principle that remains invariant under the transformations of a given group, one can derive a conserved quantity for each of the parameters of that group. During the latter half of the century, the connection between symmetries, conserved quantities, and Noether's theorem became central themes for physical developments, particularly in high-energy physics, pointing the way to new particles and new quantum numbers, such as isospin.

Beginning in the 1950s, high-energy physics enjoyed unprecedented experimental success, thanks to the construction of massive particle accelerators constructed to promote multinational collaborations. With this advanced technology in hand, researchers soon discovered an entire new "zoo" of subatomic particles. Some of these, like mesons, were heavy particles, but entirely new particles also entered the scene, including the most elementary constituents of matter, the so-called quarks. In 1964, physicists proposed a new classification scheme based on symmetry principles and quantum numbers, such as spin, isospin, parity, but also new quantum numbers, charm and strangeness. The underlying principles supporting this classification came from the theory of mathematical groups.[5]

In the 1950s, quantum field theories were still in a state of crisis due to renormalization problems, yet, by the 1970s, symmetry and renormalization principles were again playing a central role. This turnabout had much to do with the emergence of the Weinberg-Salam theory of electroweak interactions, a renormalizable theory that was invariant under symmetry transformations. As the first quantum theory unifying electromagnetic and weak interactions, it contributed much toward restoring confidence in quantum field theories during that decade. Thereafter, such theories became central in the development of high-energy physics and the search for a fundamental theory that unified electromagnetic, weak and strong interactions, namely, what gradually emerged to become today's Standard Model of Elementary Particle Physics.

Yet, symmetry was not only relevant for finding conserved quantities, it was also important for understanding what happened when symmetry broke down. The introduction of the idea of spontaneous symmetry breaking, borrowed from solid-state physics, provided quantum field theories with new insights. Physicists soon adopted this idea as a powerful instrument for explaining the origin of mass in a dynamical way, more specifically via the Higgs mechanism, a key feature in the Standard Model. Following this interpretation, current physical laws as we know them resulted from the spontaneous breaking of a supersymmetric state of the universe, which thereby gave rise to the mass and symmetries physicists study today.

RELATIVITY AND GRAVITATION

The genesis and development of the general theory of relativity[6] as the fundamental theory of gravitation and the basis for our present understanding of the cosmos at large reflects many similarities with the development of quantum physics, but also some important differences. Unlike quantum mechanics, which from the beginning drew on a wealth of new experimental data and which developed as a collaborative effort pursued by both experimentalists and theorists at various centers, general relativity emerged largely in reaction to inner-theoretical difficulties; moreover, only a handful of theorists developed this theory initially. Nevertheless, once a consensus emerged regarding basic theoretical principles and mathematical representations of relativity and quantum theory, both became foundational for physical research up to the present day. Indeed, they continue to serve as the basis for much of our understanding of the world, both in the microscopic and macroscopic realms.

Einstein completed the general theory of relativity in 1915, ten years after he first formulated the principles of special relativity. While special relativity theory emerged as a theory of the propagation of light and the dynamics of high-velocity particles, Einstein's general theory of relativity transformed the classical theory of gravitation by developing a new conceptual framework for spacetime physics. In the course of those ten years, he extended the scope of the original theory to produce what he called a generalized theory of relativity. Many physicists had difficulty understanding this shift, and some even claimed that Einstein's new theory contradicted his earlier one.

The theory of relativity was initially restricted only to uniform relative motion and the mathematical tools Einstein employed in 1905 were rather basic. Soon afterward, however, Einstein's initial theory took on an impressive mathematical formalism, thanks to the contributions of Henri Poincaré and Hermann Minkowski. Already in 1905, Poincaré recognized that the Lorentz transformations for inertial frames in special relativity form a group, soon to be called the Lorentz group, as distinguished from the so-called Galilean group underlying classical mechanics. Minkowski took a decisive step further in 1908 by creating a four-dimensional geometry on which the Lorentz group acted, thereby placing relativity theory in the context of spacetime physics. This marked an important conceptual shift in the relationship between geometry and physics. While in pre-relativistic physics, all physical phenomena took place in the fixed setting of Euclidean space, special relativity led to a four-dimensional spacetime that involved a mixing of temporal and spatial properties as reflected in the formulas of the Lorentz transformations.

Einstein had originally conceived of relativity as a fundamental theory for all of physics, yet a major challenge remained, namely how to account for Newtonian gravitational theory, which stood apart from the modern physics of fields. In Newton's theory, gravitation was an attractive action-at-a-distance force that propagated instantaneously. Newtonian gravitational attraction thus violated the principle of the constancy of the speed of light. Indeed, special relativity required that no physical interaction could exceed the speed of light. Einstein initially tried to wed gravitational force to relativity by means of a single scalar potential function defined over space, but he eventually gave up on this approach and chose instead to pursue an ambitious generalization of the original theory.

Already in 1907, he found an important clue, today known as Einstein's principle of equivalence. The simplest case of a reference frame that lies beyond the scope of the inertial frames allowed in special relativity is a frame that undergoes uniform linear acceleration with respect to these. A famous illustration of the principle of equivalence involves a thought experiment with a freely falling elevator. If those inside it cannot look outside, they have no way of discerning whether the elevator is at all accelerating or whether any force they may detect is due to the presence of a gravitational field. In recognizing the essential equivalence of these two situations, Einstein had turned a longstanding mystery, namely the exact proportionality between gravitational and inertial mass into a fundamental axiom of the theory.

The equivalence principle provided Einstein with a heuristics that allowed him to analyze gravitational phenomena in connection with accelerated frames of reference. After running into difficulties when applying his principle to rotating frames of reference, he realized in 1912 that the framework of special relativity, as first set forth by Minkowski and then elaborated by Sommerfeld and Max Laue, was too narrow to incorporate a field theory of gravitation. Minkowski based his spacetime geometry on a metric similar to the distance measure employed in Euclidean 4-space, but with the important difference that the Minkowski-metric remained invariant under Lorentz transformations. Still, the kinematic background of four-dimensional Minkowski spacetime was essentially independent from any physical processes that happened to take place in it. This aspect would fundamentally change with the advent of the general theory of relativity. In that setting, not only would the geometry of spacetime govern the motion of physical bodies but also the geometry itself became dependent on the content, distribution, and motion of matter in spacetime.

To incorporate the physical effects of gravitation, Einstein thus sought to extend the original geometrical framework of Minkowski spacetime to allow not only the inertial frames of special relativity but also virtually any reference frame whatsoever. This generalization carried with it the implication that the Minkowski-metric was merely a special case of a more complicated object, a differential form called the metric tensor, which served as the fundamental invariant in the theory of general relativity. By adopting this approach, Einstein broke entirely with the search for a scalar theory of gravitation and opted instead for a theory in which the metric tensor registers gravito-inertial effects on test bodies.

In searching for a way forward, Einstein at first worked in collaboration with his friend Marcel Grossmann. As it happened, Grossmann knew about a new theory, which seemed to offer the kinds of methods Einstein needed. Its authors were two Italian mathematicians, who called their theory the absolute differential calculus; Einstein would later dub it tensor calculus. Up until this time, he had largely taken higher mathematics for granted, but in this case, he dove deeply into the technicalities of a theory that had no immediate connection with physical intuition, which was Einstein's main forte. Many of the ingredients came together over the next three years, but what seemed to lie beyond his reach were fully generally covariant equations that would determine the metric tensor. Then, in November 1915, he suddenly found the right track, and in four short papers, he finally established the framework for his general theory of relativity.

As in the case of quantum physics, at the core stood a new foundational equation, the so-called Einstein field equation (which actually represents 10 independent equations):

$$R_{\mu\nu} - \frac{1}{2} g_{\mu\nu} R = -\frac{8\pi G}{c^4} T_{ik}.$$

To suggest very roughly what this says, one can think of the right side as the physical input, whereas the left hand side, later called the Einstein tensor, is an expression involving entities related to the spacetime geometry, one being the metric tensor $g_{\mu\nu}$. The other symbols on the left denote special types of differential invariants that arise in the tensor calculus. More specifically, on the right hand side, the energy-momentum-stress tensor $T_{\mu\nu}$ plays the role of the sources for the gravitational field. As such, it comprises all kinds of ponderable matter and fields, but also energy density, momentum flow and stresses. G and c denote familiar universal constants: G stands for the gravitational constant and c for the speed of light. In practice, relativists choose to simplify the physical input, but even then, the problem of solving these equations, i.e., determining $g_{\mu\nu}$ from them, is exceedingly difficult.

When Einstein was still searching for such field equations in late 1915, he followed his own special line of physical reasoning. Nevertheless, he was not entirely alone in this quest. In the summer of 1915, he visited Göttingen and presented a series of lectures there about his still tentative ideas on general relativity and gravitation. Einstein's approach strongly appealed to Hilbert, who hoped to combine Einstein's field theory of gravitation with Gustav Mie's theory of matter by placing both together mathematically. During this time, Hilbert worked closely with young Emmy Noether, who three years later published her fundamental theorems on invariant variational problems.

Around the same time as Einstein, Hilbert, too, found field equations for the metric tensor of general relativity, though by following a largely mathematical heuristics that relied heavily on exploring the consequences of invariant theory and variational methods. Hilbert also presented the basic assumptions of his own theory in an axiomatic form that differed from Einstein's understanding; the latter emphasized the need to make sure that Newtonian gravitational theory emerges as a classical limiting case. Yet, despite this divergence in their approaches, experts soon confirmed that Hilbert's equations were essentially equivalent to those Einstein had found. This convergence of theoretical results reflected the strong mathematical constraints placed on the field equations. In particular, they needed to be generally covariant, which meant they would assume the same form in *every* coordinate system.

By this time, theoretical physicists had come to appreciate the deep implications of Einstein's special theory of relativity, the most famous of which involved the fundamental equation linking mass and energy: $E = mc^2$. His general theory of relativity, on the other hand, was an entirely different matter; many physicists found its abstract nature discouraging. Nevertheless, this new approach to gravitational theory proved highly attractive to leading mathematicians and theoretical astronomers, like Arthur S. Eddington, Willem de Sitter, and Karl Schwarzschild. This comes as no surprise in view of the difficult and highly unfamiliar mathematics needed to understand Einstein's theory, which featured methods that had played no role in physical theorizing before 1915. What proponents immediately appreciated was the way in which general relativity overcame the inconsistency between classical gravitational theory, with its long-range action-at-a-distance forces, and special relativity, which postulated an upper bound on the propagation velocities of physical effects. Yet, the Einstein-Hilbert equations involved a major reconceptualization of the basic concepts of physics, even though this only evolved by and by in the course of elaborating on their consequences.[7]

The field equations of general relativity are both subtle and very difficult to solve, not least because of the mutual interaction between the spacetime geometry and the matter fields it contains, a feature reflected by the left and right hand sides, respectively (see above). This vexing feature might remind us of the fairy tale about Baron von Münchhausen, who found himself stuck in a swamp, but managed to get out by pulling on his own hair. In the case of general relativity, a solution to the gravitational field equations simultaneously determines the geometry that governs the motion of bodies as well as the distribution of matter in an initially undefined geometry. Once found, however, this geometrization of physics has wonderful and far-reaching consequences, as illustrated by the equations of motion. In classical mechanics, the equations that govern the motion of physical bodies under the influence of a given force field are independent of the equations that determine the field itself. In general relativity, on the other hand, the equations of motion and the field equations are no longer independent. In fact, the latter actually imply the former, even though the technical details involved turn out to be subtle.

PHYSICS AND THE WORLD AT LARGE

Soon after its inception, Einstein's general theory of relativity enjoyed some striking empirical successes. Its confirmation through three classical tests helped promote acceptance of the theory in the physics community. These involved gravitational redshift, the bending of light, and the anomalous advance of Mercury's perihelion. Since the end of the nineteenth century, astronomers had puzzled over this anomaly, but despite numerous efforts, none of them could come up with a convincing explanation for this tiny effect. Mercury orbits the sun along an elliptical path, which rotates slowly over time, displacing its perihelion (the point nearest the sun) by around some 5,600 arc seconds per century. Newtonian gravitational theory could account for most of that displacement, but astronomical findings had reached a level of precision that clearly left a discrepancy of 43 seconds of arc per century unaccounted. Einstein had also been pursuing this problem for a long time, and in November 1915, he finally cracked it by using a linearized version of his field equations to account for the missing displacement.

The second classical effect predicted by Einstein's theory also involved highly precise astronomical measurements. Years earlier, he had already queried an American astronomer about the possibility of measuring light deflection during a solar eclipse. Back then, in 1911, he held out hopes that astronomers could confirm the bending of light rays passing near the sun, thereby affirming this relativistic effect. By the time he found the field equations for general relativity, however, he realized that the effect, even though quite miniscule, should be twice as great as what he had predicted based on special relativity. This again reflects how the new role of geometry in physics was coming to the fore. In general relativity, heavy objects like our sun cause a nearby warping of space, so that light rays, which would otherwise travel along straight lines, will be bent ever so slightly by the sun's gravitational field. In 1917, British scientists, led by Eddington, began making elaborate plans to test Einstein's prediction during the solar eclipse that would take place two years later, in May 1919. To optimize the sightings and correlate measurements, teams of researchers traveled to two remote locations; one went to Sobral in northeastern Brazil, the other to the island of Principe off the western coast of Central Africa. In November of that year, British scientists were able to announce the confirmation

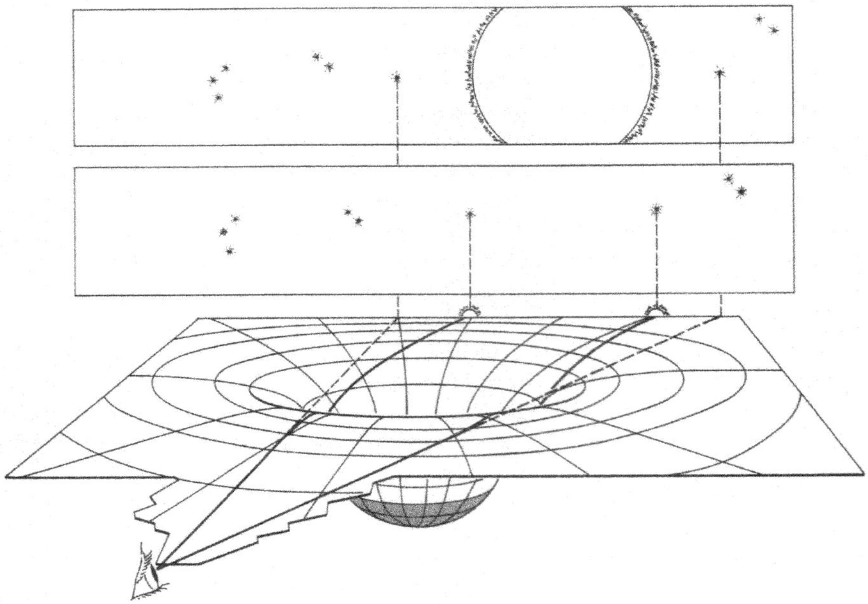

FIGURE 5.5. A schematic illustration of light deflection in the neighborhood of the gravitational field of a massive astrophysical object. Adapted from: C. S. Misner, K. S. Thorne, J. A. Wheeler. *Gravitation*, Freeman and Co., San Francisco, 1970, p. 32.

of Einstein's prediction, a dramatic event that made the then little-known German physicist into a famous celebrity almost overnight.

No such drama accompanied the third classical test, as many technical difficulties stood in the way of detecting gravitational redshift. Unequivocal confirmation of this prediction had to wait until the 1950s, when astronomers first observed this phenomenon for infrared light by making use of the newly discovered Mössbauer effect.

By now, a fourth test of Einstein's general theory of relativity has also come to be included among the so-called classical tests. This concerns the phenomenon of gravitational waves. The expectation that a relativistic theory of gravitation would imply the propagation of disturbances of the gravitational field as waves, just as Maxwellian electrodynamic field theory implied electromagnetic waves, had been a part of Einstein's heuristics in his search for a relativistic theory of gravitation. In fact, just a few months after the publication of his field equations, Einstein turned to an investigation of this question and predicted the existence of gravitational waves, another tiny but conceptually fundamental consequence of the new theory.

Yet, the possibility of confirming that prediction faced nearly insuperable difficulties, beginning with the problem of developing a theoretical treatment of astrophysical models that would point the way to the possibility of observing gravitational waves. A much more serious difficulty, however, involved building the equipment required to carry out such a test. This involved isolating a long laser interferometeric apparatus from all kinds of potential physical disturbances—seismic, mechanical, electromagnetic, and other vibrations—so that the minute distortions of the lengths of the interferometer arms due to gravitational waves might be detectable. Such an undertaking required a multi-institutional and multinational effort, but in 2015, almost exactly 100 years after Einstein first wrote about them, researchers finally detected and measured gravitational waves using large highly sensitive interferometers. In recognition of this feat, the leaders of this long-term project earned the Nobel Prize in Physics for the year 2017.

Other major implications of the gravitational field equations would emerge during the decades that followed their publication in 1915. This included the very difficult problem of finding explicit exact solutions of the field equations, beginning with the most famous case, published by Karl Schwarzschild only shortly before his death in May 1916. The Schwarzschild solution concerns the very special and highly symmetric gravitational field associated with a static, spherically symmetric mass distribution. Another early solution involved a similar static distribution of electrically charged masses, but the exact solution for a rotating spherical mass had to wait several decades before Roy Kerr obtained it in 1963.

Yet, even the simple case of the Schwarzschild solution carried profound consequences for our understanding of space, time, and causality. Only decades later, however, did experts come to realize these implications fully, by which time advances in astrophysics were paving the way to a vastly different cosmological picture. By around 1960, general relativists understood that the Schwarzschild solution implied the existence of a spacetime singularity at its origin. Moreover, the so-called Schwarzschild radius that seemed to separate an inner region surrounding the origin from its exterior was not, as had long been thought, a true boundary for spacetime events but rather a spacetime horizon. This meant that an outside observer could never see beyond it to obtain information about what goes on inside the Schwarzschild horizon. In other words, once light or matter enters this region, it can never come out again. This very feature gave rise to the popular name for the Schwarzschild solution: it represents a so-called "black hole."

The reality of such black holes in the universe has long been a point of dispute, which led to a good deal of astrophysical theorizing and modeling of stellar and galactic collapse. Nowadays, it is widely accepted that each galaxy has at least one black hole at its center. Recently, astrophysicists have produced evidence of a black hole in our own Milky Way, captured both by photographing light phenomena near the horizon and by tracking the orbits of stars close to the black hole. In fact, in 2020 another Nobel Prize went to the researchers who detected this supermassive black hole. The numerical simulation of Schwarzschild and Kerr black holes, on the other hand, was at the core of theoretical efforts to model astrophysical scenarios that would lead to the emission of gravitational waves on a measurable scale. A particular scenario involved the relative motion of two smaller black holes that eventually merge into a single large black hole. Such an astrophysical catastrophe would release enormous amounts of energy, including the emission of gravitational waves that might still be detectable even if the event took place many billions of light years away.

When the theory of general relativity first emerged, the known or even imagined universe was much tinier than today. Nevertheless, already in 1917, Einstein published an important paper that launched the new field known as relativistic cosmology. This marks the beginning of a fascinating chapter in the history of twentieth century astronomy and cosmology, a field shaped by an ongoing rapprochement between observational data and theoretical speculation. In retrospect, an important episode took place in 1920, when the American astronomers Harlow Shapley and Heber Curtis engaged in their "Great Debate." At the time, they were merely continuing a whole series of longstanding debates about the mysterious celestial phenomena astronomers called nebulae. In this case, though, the question at hand was whether these objects resided within our own galaxy, as Shapley claimed, or beyond it, the position taken by Curtis. As it turned out, the latter was on the right track, though it would take another decade to confirm that hunch. Determining distances for astrophysical objects, rather than their angular position, is a difficult problem that involves the interplay of astrophysical modeling and observation.

Speculation that objects like the Andromeda Nebula might themselves be distant galaxies was nothing new, and those who forwarded this notion realized that, if confirmed, it meant the observable cosmos would be many orders of magnitude larger than the Milky Way. The astronomer Edwin Hubble clinched the matter by his observations of Cepheid variable stars in spiral nebulae, which indicated distances far beyond the bounds of our galaxy. Hubble made another discovery, however, that proved even more crucial, even though others had observed it before him: several nebulae of interest displayed a systematic redshift, which meant that these distant objects were receding from us. In making this observation quantitatively precise, Hubble established a law asserting that the velocities by which galaxies recede from us increases in proportion to their distance. While Hubble's law may suggest at first sight an anthropocentric cosmology, it actually expresses the property of an expanding universe, one in which the metric tensor increases with time, as illustrated by the distance of dots painted on a balloon that is blown larger and larger.

Edwin Hubble's observational discovery of receding galaxies led to a flurry of interest in cosmological models based on solutions of Einstein's equations that describe an expanding universe. Einstein himself had neglected to consider this possibility, though he was aware that the Russian Aleksandr Friedmann had exhibited such a solution of the field equations in 1922. Five years later, the Belgian theorist Georges Lemaître also anticipated Hubble's findings, another indication of awareness that Einstein's field equations were highly compatible with models of an expanding universe. (Einstein, in fact, felt compelled to modify them in 1917 by adding a "cosmological term" in order to obtain a static solution.)

Extrapolating the expansion backwards in time suggests the concept of a singular origin of the temporal development of our universe, a doctrine that came to dominate our present understanding of a big-bang cosmogony. Dynamical details of this cosmological expansion, including the question whether the rate of expansion itself is changing, as well as speculations about a period of rapid inflationary cosmological expansion, are current topics of cosmological and astrophysical research. All of these investigations, whether theoretical or observational, involve models based on solutions of Einstein's gravitational field equations.

In the first decade following the inception of general relativity, many of those who worked on the theory hoped it would lead even further to a grand synthesis. They thus searched for an over-arching mathematical framework that would combine both the electromagnetic and gravitational fields into a unified field theory. These developments involved a good deal of interplay with new mathematical theories, such as multidimensional differential geometry. With the success of quantum mechanics in the late twenties, however, theoretical interest in general relativity abated somewhat, leading to a sterile period with little activity, a low-water mark that only ended in the late fifties and early sixties. Major breakthroughs in astrophysical observations, such as the discovery of pulsars in the early sixties, along with renewed attempts to understand astrophysical observations and cosmological models based on Einstein's theory, led to a renaissance of general relativity that continues to the present day. Three major events of that renaissance were the first identification of a gravitational lens in 1979, the observation and explanation of a neutron star binary pulsar by Hulse and Taylor, and finally the detection of gravitational waves in 2015.

As we have seen, quantum physics and relativity theory have transformed our understanding of the world, both small and large, in unexpected ways. Experimental physicists and astronomers have also produced a great deal of evidence that confirm both theories in many ways. A key development in quantum physics came with Dirac's equation, which made Schrödinger's equation compatible with the requirements of special relativity. It also pointed the way to quantum field theory and the Standard Model of Elementary Particle Physics. On the other hand, the search for a way to harmonize quantum physics and the general theory of relativity continues, and the problem of finding a satisfactory theory of quantum gravity remains one of the major challenges of theoretical physics today.

CHAPTER SIX

Mathematics and Technological Change

"Programming Mathematics"

LIESBETH DE MOL

In 1976, the historian of mathematics, Kenneth O. May delivered a keynote address at an important conference on the history of computers held in Los Alamos. This major *international* event brought together renowned figures from the field of computing with the aim of gathering their thoughts about recent developments. It also hoped to raise awareness of the significance of the actors own personal contributions and histories. At the time, leaders from the computing field were still looking toward mathematics and formal logic as a possible foundation for their own field. It was from this perspective that the conference organizers wanted to hear the views of a historian of mathematics (see De Mol and Bullynck 2018). One of the purposes of May's talk, entitled, "Historiography: A Perspective for Computer Scientists,"[1] was to emphasize the significance of the computer for the history of mathematics. Beyond that, he also hoped to integrate the history of computing into the latter subject.

Interestingly, May did not view the computer as just another technology belonging to the field of so-called applied mathematics. Quite the contrary, in his understanding, "[t]o say that [. . .] [the computer is] one of the most important inventions in mathematics or in the history of mathematics would be an understatement."[2] This fitted with a view that rejected the notion of a sharp dichotomy between applied and pure mathematics. According to May, it was a mythic construction to claim that some part of mathematics was disconnected from the world and purely theoretical, whereas another part was only concerned with applications of theory. This viewpoint went hand in hand with May's Marxist convictions. He proposed another distinction instead, namely, that between mathematical science and mathematical technology. The latter included an algorithmic tradition (which he called the software side of mathematical technology) but also a tradition concerned with calculating devices (the hardware side). In May's view, the advent of the electronic computer had enabled mathematical technology to assume its rightful place in the history of mathematics. As he put it:

> [Mathematical technology] developed steadily in history without an elaborate literature, so that it doesn't appear in the historical record as a central thing. [. . .] If we look at the history of mathematics this way, it seems to me that one of the effects of the coming of the electronic computer is that for the first time mathematical technology has become a consciously recognized discipline [. . .] instead of just being for practitioners who didn't seem to be in the mainstream, computing now becomes the Queen of Technology.[2]

May's remarks moved in two directions, one aimed at promoting awareness among computer scientists of their field's place in the history of mathematics (unsuccessfully, though, see the final section), another at broadening the horizons of historians of mathematics to encompass mathematical technology and computing. The principal aim of this chapter is to focus on this second direction in order to evaluate the role of modern computing as a mathematical technology: how do the histories of computing and mathematics intertwine or not? What does this tell us about perceptions of the history of mathematics? How do its practitioners regard these matters today? In order to give focus to this chapter, I will concentrate mainly on the United States. This does not mean, of course, that one should overlook relevant histories in other national and cultural contexts,[3] but these lie beyond the scope of the current chapter.

In what follows, I will discuss four major topics in as many sections: the early history of computers in relation to mathematics; the use of computational technology within mathematics; mathematics as a tool for computation; and the changing disciplinary identity of mathematics from the perspective of computational technologies.

TABLES AND OTHER AIDS TO COMPUTATION

Let me begin by asserting that one cannot reduce the prehistory of modern computing to mathematics alone, as it requires a much broader historical embedding. Amongst other factors, the industrial revolution resulted in a number of basic societal reorganizations, including the rise of formal bureaucracies that prepared the ground for the computerization of modern societies or, as some have called it, the information society (Agar 2003; Campbell-Kelly et al. 2014). Mathematics, especially so-called applied mathematics, was, however, an important driving force for the actual development of early large-scale computing technologies. In a sense, mathematics was the *raîson d'être* for many of these technologies.

Major impetus behind the push to increase computational power came from the military sciences (see also Chapter 2) and with them the realization of the potential of mathematical research to develop and deploy ever more sophisticated weapons. In the United States, the significance of military research and development increased during the First World War, as witnessed by the creation of the National Research Council (NRC) in 1916. The NRC, an initiative of the National Academy of Sciences, served as an advisory board for scientific matters relevant to the state. For mathematics specifically, several influential figures were involved with military applications, including Harvard's George D. Birkhoff, Oswald Veblen from Princeton, and Forest R. Moulton of the University of Chicago (Archibald, Dumbaugh, and Kent 2014).

During the Second World War, the US mathematical community mobilized even more decisively for the war effort. This led to the establishment of the joint War Preparedness Committee, a joint creation of the American Mathematical Society and the Mathematical Association of America, and, somewhat later, the Applied Mathematics Panel (AMP). In this context, some mathematicians became convinced that modern warfare was largely mathematical in character. Warren Weaver, then Director of the Division of Natural Sciences at the Rockefeller Foundation, became head of the AMP with Mina Rees serving as his executive assistant. Its purpose was to bring mathematicians as a group more effectively into the nation's war effort. This meant mathematics in the service of the government, as many of the AMP's projects came at the request of the Army, Navy or other NRC agencies. By the end of the war, the AMP had undertaken around 200 studies, half of them solicited by the military. A major component in this context were studies directed at computational methods. Of the four main topics mentioned in a series of three AMP summary reports from 1946, one was devoted to computational services. These included such projects as making tables and charts, the evaluation of integrals in several settings (from ballistic trajectories to the shape of shells), and research into the potential of computing equipment.

A particularly important center for war-related research was the Aberdeen Proving Ground (APG), founded only a few months after the US had entered the First World War. Here, military experts designed and tested weapons, and in both World Wars, the APG played a major role in bringing mathematics and military applications closer together. In 1918, Veblen took charge of experimental ballistics and the production of firing tables used in the field for

FIGURE 6.1. A view of the ENIAC with Marilyn Wescoff and Ruth Lichterman wiring the machine. Photo by © CORBIS / Corbis / Getty Images.

aiming artillery. These tables were quite elaborate, as a number of factors influenced the trajectory of a shell, including air temperature, barometric pressure, wind velocity, as well as the gun's inclination. Firing tables contained the relevant information related to these various parameters and were employed to achieve optimal aim of the gun.

Since tables had to be compiled for every new weapon, there was a growing need for them as the military efforts of the US increased. The preparation of these firing tables at the APG—more specifically, at its Research Department, renamed the Ballistic Research Lab (BRL) in 1938—became a major contributing factor in the development of computational technologies. This pertained not just to the technology itself but also with respect to the development of improved mathematical methods (Aspray 1990). Gradually, the BRL became a hub for work with and exploration of computing technologies and computational methods, attracting a diverse group of mathematicians as consultants including, amongst others, Haskell B. Curry, John von Neumann, and Isaac Jacob Schoenberg.

At the BRL, computation of firing tables initially relied mostly on human "computers," people working with desk calculators. By the 1930s, they carried out computations using two differential analyzers (note that our current English notion of "computer" derives from humans computing). At this time, these were state-of-the-art calculating devices, as only shortly before Vannevar Bush and

Samuel H. Caldwell built the prototypes for this types of differential analyzer at the Massachusetts Institute of Technology(MIT). Today, we would characterize these as analog machines, that is, calculators that work with continuous measurements as opposed to digital machines, which work with discrete numbers. By the late 1930s, these analyzers had the capability to solve *any* differential equation within the machine's computational capacity. They were thus well suited to solve the basic differential equations involved in the computation of ballistic trajectories.

However, even with the help of these differential analyzers, the teams at BRL could not keep up with the increasing demand for firing tables. This problem came to the attention of John W. Mauchly and J. Presper Eckert from the Moore School of Electrical Engineering of the University of Pennsylvania, an institution in close contact with BRL. In 1943, Eckert and Mauchly proposed to build an electronic high-speed machine to compute firing tables, a proposal that the army soon accepted. This led to the construction of one of the first large-scale electronic computers, the ENIAC, though it took until early 1946 to complete (see Figure 6.1). Even though it was then too late to serve its initial purpose, the ENIAC proved useful for a range of other applications. These included computations in number theory (carried out by Derrick and Emma Lehmer), for weather prediction in the Meteorology Project (led by Jules Charney), and for the famous Monte Carlo methods used in the design of nuclear weapons. John von Neumann, who was deeply involved with the Manhattan project, spearheaded this latter effort (see Chapter 2). The Monte Carlo approach relied on the use of random numbers to study the pattern of simulated paths of neutrons that scatter, escape, or are absorbed during nuclear fission, in order to obtain a statistical picture of their behavior.

As will be discussed later in more detail, the introduction of high-speed discrete computation would have a major impact on the mathematical methods used for the computation of numbers. To give a very basic example, let us consider the method for computing the division of one natural number by another. A familiar method used by humans is long division. However, for long division you need to make use of educated guesses throughout the process, so this was not well suited for early computers. Other methods had to be used that were better adapted to the capabilities of the machine. Today, a whole literature exists on the efficiency of different methods for division, including for instance the SRT algorithm. This is the computer's equivalent of long division and relies on hardwired tables inside the machine. On the software level, one uses other types of methods. In general, the method chosen depends very much on the local context in which the implementation takes place, taking into account, for instance, the cost for the speed and memory required.

Yet, while military demands were certainly an important factor behind developments in computational technologies, this was only part of the story. In fact, this aspect can be embedded in a more general context involving an increased use of computational methods in a number of different scientific fields. During the first half of the twentieth century, a number of large-scale scientific projects relied on extensive computation, often with the aim of producing numerical tables, a practice with a long tradition (Campbell-Kelly et al. 2003; Grier 2005). In these projects, one often relied on calculating machines and other technical devices that made the computational process more efficient. These devices included desk calculators, tabulators, and assemblies of smaller interconnected calculators. Leslie J. Comrie used a Brunsviga-Dupla machine for astronomical computations (see Figure 6.2), and in the 1930s founded one of the first private firms that specialized in scientific computation.

Mechanized computations often went hand-in-hand with computational work done by teams of human computers following a more or less strict division of labor. One such large project was located at the Nautical Almanac Office in the UK, where Comrie introduced desk calculators in the 1930s. Another was the Mathematical Tables Project in New York, led by Arnold Lowan and Gertrude Blanch, which at times employed more than one hundred human computers. Yet another major effort involved computations for astronomical research, work undertaken at the Astronomical Computing Bureau under the direction of Wallace J. Eckert. This later became the Watson Scientific Laboratory for Astronomical Computations, recognizing financial support from IBM and Thomas J. Watson, in particular.

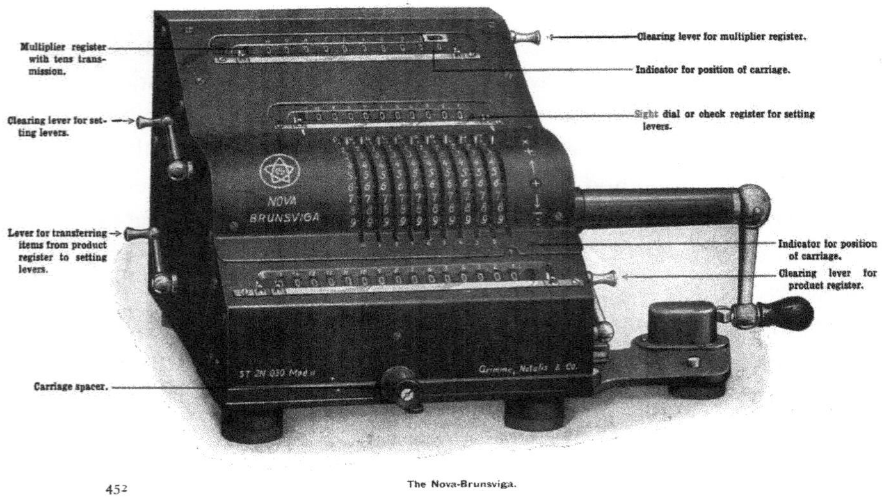

FIGURE 6.2. The Brunsviga-Dupla calculator as presented by Comrie in 1928 before the Royal Astronomical Society in 1928. Image courtesy of Liesbeth De Mol.

The first large-scale machines were initially conceived as automating the work of human computers. For instance, as Maurice Wilkes pointed out in the context of the EDSAC, an early computer constructed in Cambridge, England:

> With the EDSAC the analogy between the machine and a computor is a close one in that the problem is presented to the machine in the form of a series of orders [. . .] In a metaphoric sense the machine may be said to understand the same language as a computer.[4]

These organized practices utilizing human workers helped to establish computing as a valuable scientific method. This image gained support through the creation of organizations like the Mathematical Tables Committee in the UK and its counterpart in the US, the National Research Council Committee on the Bibliography of Mathematical Tables and other Aids to Computation. Raymond C. Archibald from Brown University, a leading figure in the American Mathematical Society, headed the latter committee (Grier 2005).

In 1943, Archibald created the journal *Mathematical Tables and Other Aids to Computation* (MTAC) with the purpose of reporting on recent developments in the production of mathematical tables and related computational techniques. Soon thereafter, in the late 1940s, it also played a key role in reporting on work done in relation to the new large-scale computing machines then being constructed. It also became the recommended journal of the Eastern Association for Computing Machinery (EACM), later the ACM,[5] before that organization founded its own journal in 1954. By the 1950s, however, interest in the MTAC had declined. Ironically, this was at least partially due to the rise of the modern computer. Precomputed values in mathematical tables were fast becoming obsolete since, using these computers, one could quickly compute values as needed (see below), which meant that attention shifted more to the mathematics of computation itself. In 1959, the journal MTAC dropped its name and became the *Mathematics of Computation*, which today covers topics such as numerical analysis and computational discrete mathematics.

Despite the fact that the relevance of mathematical tables as human calculating aids steadily declined with the rise of high-speed computers, many of those involved in table making became pioneers in modern computing. Archibald, through the establishment of MTAC, was one; another was Derrick H. Lehmer, one of the first ENIAC users (see Introduction). Lehmer learned about number theory from his father, who taught him that the subject requires exploration and, by consequence, extensive computation. To this end, the younger Lehmer began constructing several special-purpose devices to help in the computation of tables of prime numbers (so-called prime sieves), and he soon became a specialist in number-theoretical tables. In 1943, he became a member of the editorial board of MTAC and, one year later, its co-editor with

Archibald. During the Second World War, Lehmer became involved with research for the Applied Mathematics Panel, which led to the construction of a special-purpose device to "simulate" random-bombing operations. Probably through this work he became a member of the Army Computations Committee that tested the newly built ENIAC and other machines. Later, Lehmer would be an important advocate of so-called computer-assisted mathematics (see below).

As a graduate student in electrical engineering at Harvard, Howard Aiken came to recognize that extensive computations were essential to solve the differential equations that describe the behavior of electrons in vacuum tubes. Harvard's astronomical observatory had long hired human computers to deal with such problems, but Aiken believed that such help would not suffice for handling such a complicated case. Instead, he thought it necessary to employ a machine that could replace human calculators. IBM decided to help construct such a machine, known as the ASCC, short for Automatic Sequence Controlled Calculator. This referred to the fact that, unlike desk calculators, it could carry out automatically a sequence of (coded) mathematical operations, which were "read" by the machine from a punched tape. The ASCC, later known as the Harvard Mark I, represents a landmark in the early history of computers. Its nickname was Bessie because it was used, among other purposes, to compute tables of Bessel functions (some fifty volumes of Bessel functions were published). The US Navy also used the Mark I in a program under the leadership of Grace Hopper, who took her PhD in mathematics from Yale in 1934. In the 1950s, when working with UNIVAC, Hopper became a major contributor to automatic programming in connection with the development of the programming language COBOL. The UNIVAC series of commercial electronic computers was used for compiling census data, but most famously, in predicting Eisenhower's landslide victory in the 1952 presidential election.

One should note that women had long been engaged to work on larger mathematical and computational projects. Grace Hopper was perhaps the best-known expert of her generation, but there were others, like Mina Rees, who later served as head of the mathematics department of the Office of Naval Research. Gertrude Blanch led the *Mathematical Tables Project* and several women worked with ENIAC, including Adèle Goldstine, Jean Bartik and Wila Wyatt. This employment picture was in part due to the new opportunities that opened during the Second World War but also to the fact that women had long filled lower-paying jobs as human computers.

The emergence of the modern computer was not merely the result of an effort to mobilize mathematicians for the war effort. Computational methods had already before made important advances in science, administration and engineering. Thus, when the US entered the Second World War, its leaders could draw on this know-how as background for exploring and attacking new problems that often demanded large-scale computations. In this context,

FIGURE 6.3. Grace Hopper, one of several women who were involved with early computational practices, working at a terminal of the IBM ASCC / Harvard Mark I machine. Photo by Bettmann / Getty.

mathematicians needed not only solid training in abstract mathematics but also a willingness to engage with computing machinery and engineering technology. Many of those who rose to the challenge became pioneers in the field of computing. Working alongside engineers and operators, they contributed to shaping the field of electronic computing, its basic problems and, later, the identity of the disciplines that emerged from these efforts.

DOING MATHEMATICS WITH THE COMPUTER

As noted above, mathematics was not only an important factor in the construction and use of the first computers; these machines, in turn, had a profound impact on mathematics itself. Indeed, high-speed computation makes it possible to study problems that mathematicians could never solve by traditional methods. This has tremendously widened the range of phenomena researchers can study by mathematical means, a trend that also led to an entirely new field of research: today's experimental, explorative or quasi-heuristic mathematics. The apparent opposition between traditional mathematics, on the one hand, which many view as being about proofs and theorems, and experimentation, on the other hand, has often haunted the field of so-called "experimental mathematics" and given rise to ongoing controversies between advocates and opponents.

Many real-world phenomena have a complexity that far exceeds the capabilities of traditional methods for problem solving by mathematical analysis. Such phenomena include, for instance, studying nuclear reactions, atmospheric turbulence, or pattern bombing (Galison 1997). Even before the advent of the first high-speed computers, problems such as these led to innovative methods based on number crunching. For instance, in connection with the Manhattan project, von Neumann studied phenomena like implosion and shock waves using calculating machines such as the Harvard Mark I. Another example relates to a military problem studied by mathematicians associated with the AMP, namely, to determine the probability of hitting a certain area with pattern bombing (the simultaneous release of bombs carried by a formation of aircraft). They were handled by means of so-called "model experiments," which used randomized methods to produce statistical models.

Although these efforts to attack difficult problems pointed in a new direction, their success was limited. In relying on innovative methods and mechanical computational aids, mathematicians encountered difficulties simply due to the time required to carry out so many calculations. Once high-speed electronic computers became available, they could achieve far more, especially if an entire set of problems could be set-up on a single machine. As several advocates and pioneers of computer-assisted mathematics emphasized, one should not imagine that increasing the speed of computation by several orders of magnitudes merely represented a quantitative improvement. That would be like saying that the change brought about by modern automobiles and airplanes led to a reduction in our walking time.[6] Moreover, these high-speed computers were also programmable; this meant they could take on problems with a long or complex program structure, like those involving Monte Carlo computations. Mathematicians could program computers to generate random numbers, thereby obviating the need to rely on tables.

In earlier times, one would have considered the types of problems discussed above as falling within applied mathematics or mathematical physics to the extent these problems were even amenable to conventional mathematical methods. In today's world, problems such as these have led to entirely new approaches based on computer simulation and modeling (in climate science, for instance). Within this context, mathematics has become a tool used by specialists in another field. These developments have not only raised interest in computational mathematics they have also rejuvenated older mathematical theories in cases where computers have made it possible to revisit those ideas.

A particularly striking example of this can be seen in research on non-linear dynamics and chaos (Aubin and Dahan Dalmedico 2002)—a field first revived in the wake of work undertaken by Stanislaw Ulam, John Pasta, Enrico Fermi, von Neumann, and Charney. Non-linear phenomena are notoriously difficult to study because, practically speaking, they are nearly impossible to predict. A familiar example of such non-linear behavior comes from weather systems, which evolve under constantly changing conditions (for examples of non-linear system in physics, see Chapter 5). Even very simple non-linear mathematical processes can easily give rise to highly complex behavior. As a classic example of this, Ulam and von Neumann introduced and studied a special case known as the logistic map. Originally, they studied this in the context of research on algorithms for computing so-called pseudo-random numbers, that is, values computed by a deterministic and algorithmic procedure, but which statistically appear to be random.

THE LOGISTIC MAP

The logistic map is a recursive function where the output of one iteration becomes the input of the next, as shown in the flow diagram in Figure 6.4.

Ulam, Pasta and Fermi began exploring non-linear dynamics with computational methods in 1955 using the MANIAC computer in Los Alamos, around the same time that Charney was using the ENIAC at the Meteorology Project. Charney's work on long-term predictability paved the way for studies in the early 1960s undertaken by Edward Lorenz, a mathematician and meteorologist. Lorenz showed that such weather predictions were practically impossible due to the sensitivity to the initial conditions, today known as the

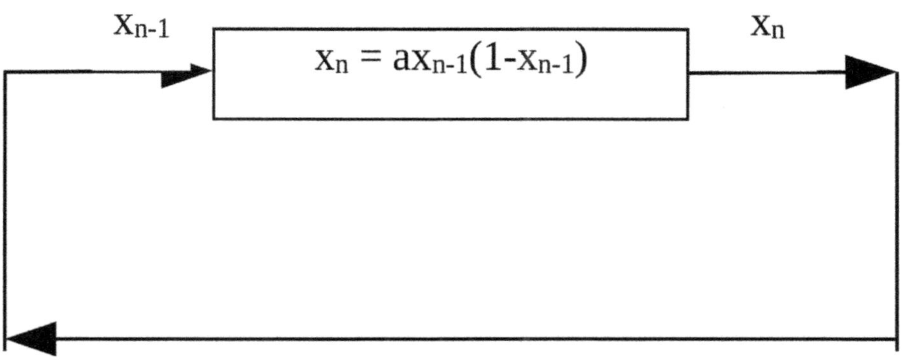

FIGURE 6.4. The value of x is a real number between 0 and 1, the more interesting cases are those where the parameter *a* is between 0 and 4. When a = 4 the map is known to show so-called chaotic behavior. The map is known as one of the simplest possible examples of a mathematical iteration that shows complex behavior. Image courtesy of Liesbeth De Mol.

butterfly effect. The idea is that a tiny atmospheric change, caused perhaps by a butterfly flapping its wings, can lead to large-scale effects on the weather. This example serves to underscore how even in physical systems presumed to be completely deterministic (thus devoid of quantum effects), their evolution may still be inherently unpredictable. Computers have nevertheless long played a key role in meteorology, both through extensive computations as well as visualizations of weather patterns. A well-known theoretical visualization involves the so-called Lorenz attractor, as shown in Figure 6.5.

The first high-speed computers were employed, however, not only to attack physical problems (e.g., those related to nuclear fission) but also to study classical problems in mathematics proper, particularly in the field of number theory. These investigations continued work from an older tradition that used computations to test open conjectures, like the Riemann hypothesis, Fermat's

FIGURE 6.5. Visualization of the Lorenz attractor. It shows the well-known butterfly effect in weather prediction showing the sensitivity to initial conditions in which a small change at an earlier stage can have large changes in weather at a later stage. The butterfly effect then is a metaphor for this saying that distant flapping of the wings of a butterfly can have basic effects on the weather later. The attractor shows that phenomenon. Photo by Science Photo Library / Getty.

last theorem or Goldbach's conjecture, or simply to explore properties of certain classes of numbers. In the era of high-speed computers, perhaps the earliest studies were those done on ENIAC by Derrick and Emma Lehmer. These concerned the computation of tables relevant for identifying prime numbers. During this initial stage, no one knew whether a high-speed computer could really be useful for doing something as sophisticated as number theory. In a parallel development in Britain, Alan Turing continued his earlier work on the Riemann hypothesis, for which he had already constructed special purpose devices on the Manchester Mark I. Turing then thought that by computing more and more values of the Riemann zeta function he might find a counterexample that would invalidate the conjecture. Lehmer did similar work on ENIAC and later on SWAC, which the Lehmers and Harry Vandiver also used for work on Fermat's conjecture (Corry 2008a).

A recurring approach, one that continues to this day, made use of the idle time on a computer to compute more and more values of a given number-theoretical function. Programmers thus utilized the idle time on ENIAC to compute decimal expansions for π and e—considered a possible source of random numbers by von Neumann—but also Fermat quotients, which are relevant for Fermat's last theorem. They used a versatile program called "Slow Moses," which could not only do calculations in number theory but also at the same time test the stability of the ENIAC hardware. In the UK, mathematicians used the idle time on EDSAC to compute Mersenne primes with a program later run on the Pilot ACE, which was faster than EDSAC. A Mersenne prime P_n is a prime number of the form $2^n - 1$; these are very hard to find by hand, as the computations are extremely laborious and error-prone, though not requiring much mathematical ingenuity. It is unknown to this day whether there are infinitely many Mersenne primes; if not, there would have to be a largest such P_n.

As these few examples illustrate, computers were used—continuing an older tradition in number theory—to compute large values for special functions. One hoped that these results might lead to counterexamples or perhaps provide new insights into existing problems. The Mersenne prime program is one of several such projects that continue up to the present today. The so-called Great Internet Mersenne Prime Search (GIMPS) is an interesting offshoot stemming from it. Since its founding in 1996, participants have discovered sixteen new Mersenne primes, the largest of which has a length of nearly 25 million digits. GIMPS is an example of a large-scale distributed computing project—a project where a large number of connected computers work together. Anyone who installs the relevant software can contribute by using their computer's idle time. As was the case with the "Slow Moses" program on ENIAC, this project not only served number theory, though its larger significance stems from having demonstrated the potential of distributed computing.

Numerous other examples of the role of computers in research in number theory might be mentioned, like the search for exceptions to the Collatz conjecture (also known as the $3n+1$ problem), or research on prime gaps and prime constellations (e.g., twin primes that differ only by 2). These efforts are hardly restricted to tabulating cases; many employ sophisticated methods that involve looking for patterns. Finding these can help to determine heuristically certain properties, which can then assist in proving conjectures or formulating new ones. Interestingly, these kind of researches are not just relevant for number theory alone. As a side effect, they have resulted in new algorithms or insights into existing hardware and software. A relatively recent example came about when number theorist Thomas Niceley discovered the FDIV Pentium bug, a glitch that cost Intel millions of dollars. Niceley had noticed an inconsistency in his results while doing research on prime gaps. This then led him to realize there was an error on the hardware level (there was an error in the hardwired table of the SRT division algorithm).

Just as advances in hardware and software have affected the way people today do mathematics, so has the internet. A striking example is the *Online Encyclopedia for Integer Sequences* (OEIS: https://oeis.org/), initiated by Neil Sloane. This explorative tool, which currently contains more than 300,000 integer sequences, has been used and cited in literally thousands of mathematical papers.[7] One of the interesting features of OEIS is an algorithm called Superseeker. If you submit a sequence of integers to Superseeker, it will attempt to find an explanation for your sequence by relying on OEIS combined with an extensive library of programs. To give a simple example, suppose you submit the following sequence of numbers:

2, 3, 5, 7, 11, 13, 17, 19, 23, 29, 31, 37, 41, 43, 47, 53, 59, 61, 67, 71, 73, 79, 83, 89, 97, 101, 103, 107, 109, 113, 127, 131, 137, 139, 149, 151

No doubt, Superseeker would answer that this sequence lists the first 36 prime numbers. But what would happen if you sent in a sequence that you chose seemingly at random?

Beyond the realm of natural numbers, computers have led to new insights for other classical problems. An impressive example was the "discovery" of the Bailey-Borwein-Plouffe formula, which resulted in an improved algorithm for computing π. As the authors explained, they came upon this not by formal reasoning but rather by a combination of inspired guesswork and extensive searching using the so-called PSLQ algorithm, which finds integer relations in sets of real numbers.

Computers, to be sure, are important in pure mathematics not only for establishing experimental results but also for proving theorems, especially when these involve a large number of cases. Such computer-assisted proofs have the capability of examining vast numbers of particular cases, which need to be determined and then verified, relying on a complex of computer programs. Probably the most controversial and best-known example was the original proof of the four-color theorem produced by Kenneth Appel and Wolfgang Haken. This settled a longstanding conjecture, namely, to decide whether any map could be colored using only four colors in such a way that no adjacent regions have the same color.

For a variety of reasons, many members of the mathematical community did not consider this result, first published in 1976, to be a real proof (MacKenzie 2004). Amongst others, its length and the complexity of the programs involved not only made the proof intractable but it also raised a plethora of other issues. How, they asked, could one ever guarantee there were no errors due to programming mistakes or hardware failures? Moreover, critics saw this as an extreme case of a proof that provides no mathematical understanding as to why the theorem is true, simply because no human being could ever comprehend it in all details. Another more recent case came up when Thomas Hales solved the sphere-packing problem, also known as Kepler's problem. Simply put, the problem Hales solved was that of determining the most efficient way to arrange (identical) spheres in a three-dimensional space. It appeared in *Annals of Mathematics* in 2005, after a review process by a team of reviewers who worked on it for several years. Even then, these experts concluded that they were only 99 percent certain the proof did not contain errors.

The uses of the computer in applied and pure mathematics discussed here share one basic feature, namely the apparent experimental, quasi-heuristic nature of the methods and results involved. The practitioners involved realized this, of course, and they emphasized it on several occasions, sometimes in a quite provocative manner that invited disagreement and controversy. Discussions about the experimental nature of computer simulations are well known, but they also arose in pure mathematics, where experiments, exploration and heuristics play a role, even in the case of the computer-assisted proofs.

An outspoken figure during the earlier years was Derrick H. Lehmer, whose research interests long focused on concrete problems in number theory. On several public occasions, Lehmer emphasized the significance of the computer for this kind of mathematics, drawing a sharp contrast between what he called explorative mathematics and the more abstract approaches that were then currently fashionable. He was perhaps most explicit about this in 1965, when he received the prestigious Gibbs lectureship awarded by the AMS. In his lecture, delivered during the heyday of Bourbaki (see Chapter 3), he remarked:

> The most popular school now-a-days favors the extension of existing methods of proof to more general situations. This procedure tends to weaken

> hypothesis rather than to strengthen conclusions. It favors the proliferation of existence theorems and is psychologically comforting in that one is less likely to run across theorems one cannot prove. Under this regime mathematics would become an expanding universe of generality and abstraction, spreading out over a multi-dimensional featureless landscape in which every stone becomes a nugget by definition. Fortunately, there is a second school of thought. This school favors exploration as a means of discovery.[8]

This rhetorical dissonance continued to echo decades later, though in a very different climate in which the Bourbakists had all but vanished from the scene. Nevertheless, advocates of experimental mathematics had to fend with traditionalists who regarded their work as a lower form of mathematics because of its lack of rigor and "real" proofs. Many were averse to accepting results that depended on probabilistic arguments, and probably even more distrusted proofs based on complex programs. Féjès Toth, an important contributor to the sphere-packing problem and one of the reviewers of Hales' proof, compared computer-assisted proofs to findings in the experimental sciences, where referees typically cannot guarantee the correctness of an experiment.

Some advocates of experimental mathematics have embraced their role as radicals warring against the establishment. Doron Zeilberger has been perhaps the most outspoken advocate of so-called semi-rigorous mathematics that relies on probabilistic truths. Yet even though experimental mathematics has gained growing acceptance in recent decades, it still occasions sharp debates. The combatants often adopt one of two opposing views on mathematics: the older one insists that mathematical knowledge must be grounded on rigorous formal proof, whereas those who dismiss this position believe that mathematics goes beyond formal truths, while claiming that mathematical methods and results are far closer to those of the sciences.

Interestingly and perhaps ironically, it was precisely in the context of such debates that Thomas Hales embarked on the Flyspeck project. Given the many uncertainties that surrounded his proof of the sphere-packing problem, Hales decided to provide a new kind of formal proof. This proof, too, was computer-assisted, but with the novelty that a computer-assisted proof assistant subjected it to formal verification, an approach reminiscent of Hilbert's proof theory (see Chapter 4). In the meantime, the work of Appel and Haken on the four-color theorem gained confirmation by means of a similar formal proof. These kinds of projects led to important collaborations between computer scientists, logicians and mathematicians, who together designed software used in formal proofs but also in testing programming tools. A particularly noteworthy example of the latter is the CompCert project, used for the formal verification of the C compiler (the C compiler is perhaps the most basic compiler for programming languages). As described in the section following, research into the formal verification of programs forms an important chapter in the history of programming and led in the 1980s to the formal correctness debates.

Today, computer-assisted mathematics has achieved far broader acceptance than was the case during the last century, and these possibilities have continued to grow with increases in memory and speed. This process went hand in hand with easier access to computers and improved peripherals (which made better visualizations possible). Over a relatively short period of time, the computer evolved into something much more than a tool for computation—it became a heuristic tool that not only provided data but also did (and does) much of the analysis on that data by providing visualizations, graphics, etc. This has resulted in important cross-fertilizations between mathematics and the computing field, as exemplified by the Flyspeck and GIMPS projects or the discovery of the Pentium bug. Moreover, the development of software packages like Maple and Mathematica has contributed significantly to the popularization of computer-assisted mathematics. Stephen Wolfram's experimental research in the 1980s and 1990s on cellular automata led him to the development of Mathematica, one of the leading software packages today for mathematical research, alongside Maple. These products, however, have also raised the problem of accessibility. Both Maple and Mathematica are closed and proprietary, which means that the many programs written with these software packages are not accessible to the mathematicians who produce and use them. There are some open-source alternatives, like Sage, but Maple and Mathematica have remained standard.

MATHEMATICS OF COMPUTATION

Mathematics was and is not only a field of application for high-speed computation. It is also a basic field for developing new or improved algorithmic and programming methods to gain higher efficiency in speed and memory and to improve programming tools. Moreover, once it became clear that high-speed computation gives rise to a number of theoretical problems that were there to stay, a new field emerged that aimed to *study* computation and the computer as a topic in and of itself. In that sense, mathematics became not only a *telos* for computations that produce new mathematical insights, it also shapes the way computations are done (as methods) and how the results can be understood (as theory). This meant that the computer was no longer just a tool for doing mathematics; mathematics itself was also a tool for the computer. It is thus not surprising that as computers became more broadly available, the role of mathematics for computing did not remain restricted to mathematics departments at universities but spread to other fields (most notably, the new field of computer science). Moreover, mathematics also found its way more easily into business and industry largely through computers. As such, the mathematics of computation often operated along boundaries between fields—mathematics, computer science and engineering—making it impossible to draw strict borderlines between the disciplines that engaged in this work.

The historical process that led to the emergence of modern computers involved a mixture of older methods with new approaches. This development forms a chapter within the history of automation, as part of a broader transition in which machines steadily replaced human workers whose daily jobs involved performing computations. This technological shift in the workplace raised a fundamental philosophical question, as it were, namely: what and how can a process be automated? Especially during the early years of this transition, several pioneers pointed out that certain methods, which might work well for humans do not necessarily work well for a machine, and vice versa.

One such pioneering figure was Douglas Hartree, a UK mathematician who became involved with the construction and use of calculating machines (differential analyzers and the ENIAC, among others). Hartree also wrote one of the first historical overviews, in which he noted:

> in programming a problem for the machine, it is necessary to try to take a "machine's-eye view" of the operating instructions, that is to look at them from the point of view of the machine which can only follow them literally, without introducing anything not expressed explicitly by them, and try to foresee all the unexpected things that might occur [. . .]. It is quite difficult to put oneself in the position of doing without *any* hints, which intelligence and experience would suggest to a human computer.[9]

Beyond this general change of perspective, the specifics of a changing computing technology also required different approaches. First, the fact that the modern computer works digitally implies that one needs to state a mathematical problem in a discrete algorithmic form. This involves replacing non-discrete mathematical operations (e.g., transcendental operations like integration) by elementary arithmetical operations and explicit steps. This methodology naturally leads to a constructive and finitist approach to mathematical problems, moving towards what we now call discrete mathematics (see the Introduction).

Moreover, increased speed and memory also had an effect on the methods used and studied. An early example involved the rather swift transition from using mathematical tables to perform computations—which was the human way of working—to employing algorithms that allow one to compute the values found in tables directly. That shift reflected the circumstance that, at the time, the technology for electronic memory was lagging behind high-speed computation. This meant that, if during a computation one needed additional data, one had to rely on much slower electro-mechanical devices, like punched cards, which led to a significant slow-down in computation. Hence, it was far more efficient simply to compute values as needed, just as von Neumann did with the ENIAC Monte Carlo computations, when he replaced cards of random numbers by a pseudo-random number generator. This meant using an algorithm rather than storing numbers already generated by some external device.[10] Such cases meant deleting mathematical tables (stored data) to make room for sophisticated programming techniques—a trade-off that was quite common in the early years when storage capacity was limited.

The computer motivated the development, adaptation and improvement of a broad variety of algorithmic methods with diverse applications, ranging from cryptography to computer graphics and sound analysis. The steady evolution of computers, from a technology that was once difficult to access to a ubiquitous feature of the workplace and home, occurred so quickly that many today can still remember life before the PC revolution. Within mathematics, these explosive developments have had a major impact across the board, but especially on certain fields of research that had long been of marginal interest to the mathematical community.

A prominent example was the field of numerical analysis, which deals with approximation methods for analytical problems, such as solving differential equations that have no exact solution. Numerical analysis has a long history and was already growing in significance before the onset of high-speed computation, as witnessed by the increased need for mathematical tables, including the firing tables discussed earlier. Those particular tables required solving two basic differential equations, which proved to be an intractable problem when approached by classical methods; mathematicians had to resort to a step-by-step method that produced successively better approximate solutions.

The computer, however, opened the door to a vast new set of problems, including how to deal with rounding-off errors and error estimates. Working with discrete machines, one can only represent a real number by an approximation valid up to a certain number of decimal places. For instance, to ten decimal places, the number π is 3.141592654. If we would then multiply two 10-place numbers, the product must also be another 10-place number, which means one needs to round-off the answer, thereby producing systematic error. This well-known problem became much more acute with the rise of the electronic computer. To put it roughly, if a computer can perform 10,000 computations rather than merely 100, the potential error due to rounding becomes much more dramatic. This became clear, for instance, with Ulam's work on the logistic map (see the previous section), where small differences in the initial input grow exponentially with repeated iterations. The problem of estimating the errors that arise from rounding-off has become a classic subfield of numerical analysis.

Since many problems in computing require numerical analysis, the need for improved methods grew significantly with the steady spread of the computer in science and society. This proliferation process has accelerated, fed by dynamic changes on the technological level that have exploited the exponential increases in speed and memory. A striking instance arose from the work of Ingrid Daubechies (see the Introduction) on wavelets, which led to improved methods of image compression (wavelets underlie the jpeg compression algorithm). Another more recent example stems from the reliance on optimization techniques in the setting of so-called Deep Learning. This new AI technology, if we are to believe its promotors (especially those with commercial interests), will yet again revolutionize our world.

While mathematics has played a vital role in developing computational methods adapted to the computer, practitioners also came to realize that a large number of problems appeared to be specific to computation and programming itself. Soon, mathematicians found themselves facing a fast-emerging discipline known in the US as computer science (Mahoney 2011). Computer science grew into an established field in the 1960s when several universities created new departments; Purdue University established the first computer science department in the US in 1962. From the start, the field was haunted by debates about its disciplinary identity (see the next section), its content and scope, and even its name (e.g., "Informatique" in France or cybernetics in Russia). Anecdotal, but indicative in this context is the following quote from a letter to the editor of *Communications of the ACM*:

> It would help our profession to be widely recognized if it had a brief, definitive, and distinctive name. This should be general enough to cover a variety of subfields [. . .] but specific enough to imply that computing applications are involved. Consider the solid professional sound of such terms as "Petroleum Engineer" or "Nuclear Physicist." What can we use that will be equally clear-cut—and at least half as impressive?[11]

Despite these disciplinary debates, one can identify some clear developments that connected mathematics directly to computing. Here, we focus on two principal developments: first, the use and study of formal models for computing machinery, and second, the use of formal logic in programming.

When the first computers came on the scene, a few pioneering figures, who happened to be familiar with developments in mathematical logic, emphasized the connection between computing and earlier work on so-called decidability problems (De Mol 2018). One of the most famous of these, proposed by Hilbert in 1900 as his tenth Paris problem, concerns Diophantine equations, equations whose numerical solutions are restricted to be integers. Hilbert asked whether it was possible to find a uniform method (today, we would speak of an algorithm) that would enable one to decide whether a given Diophantine equation, such as for instance, $x^n + y^n = z^n$, has an integral solution or does not. The example just cited suggests how difficult it would be to produce such an algorithm. In the mid-seventeenth century, Pierre de Fermat stated that, for exponents with values $n > 2$, equations of this type have no integral solutions. That famous conjecture remained open until the 1990s, when Andrew Wiles finally proved Fermat's last theorem. (In 1970, Yuri Matiyasevich solved Hilbert's tenth problem by proving an algorithm for deciding whether Diophantine equations have a solution can never be found.)

A similar case, related to Hilbert's formalist program (see Chapter 4), is the decision problem for first-order logic. Here, one seeks a (uniform) finite procedure for deciding whether a given proposition in first-order logic is derivable in that logic. In the 1920s, Hilbert expressed his conviction that it should be possible to construct such a procedure. Others, including von Neumann, were more skeptical probably because they understood that, if one actually had such a procedure, this would imply the existence of a mechanical set of rules for solving *any* mathematical problem (at least in principle). In the 1920s and 1930s, three researchers independently proved that this and other decision problems were impossible: Alonzo Church (1936), Emil Post (1921), and Alan Turing (1936). In their investigations, one encounters what nowadays are called logically equivalent abstract models of computability. Assuming these models capture the intuitive notion of computability, one can then interpret the results of Church, Post and Turing as saying that the decision problem, or any other logical problem equivalent to it, is uncomputable—that is, there exists *no* algorithm for deciding every instance of the problem. Variants of this statement appear in the literature as the thesis of one of these authors, or sometimes as the Church-Turing thesis.

Particularly famous in this connection were Turing's results showing that the computability of any real number reduces to its computability by a Turing machine or, putting it another way, if a Turing machine cannot compute a number, then it is uncomputable (Hodges [1983] 2014). Turing derived the idea of these eponymous machines by an informal analysis of all the possible (human) processes involved in computing a number. This led him to formulate a basic set of operations for an abstract machine that scans a linear tape consisting of squares containing symbols. The machine is always in a given state and proceeds by first reading the symbol on the square it is scanning. Depending on its value, it moves the tape to an adjacent square (left or right), prints some symbol and goes to the next state. Most importantly, Turing also constructed a so-called universal machine: an abstract machine that is capable of simulating the operations of any other Turing machine. In the 1950s and 1960s, this was recast into a formal model of the modern computer; just like a universal Turing machine, so went the argument, the computer is also able to compute anything that is computable. Moreover, once a certain number of primitive operations becomes available, one can carry out any computation whatsoever.

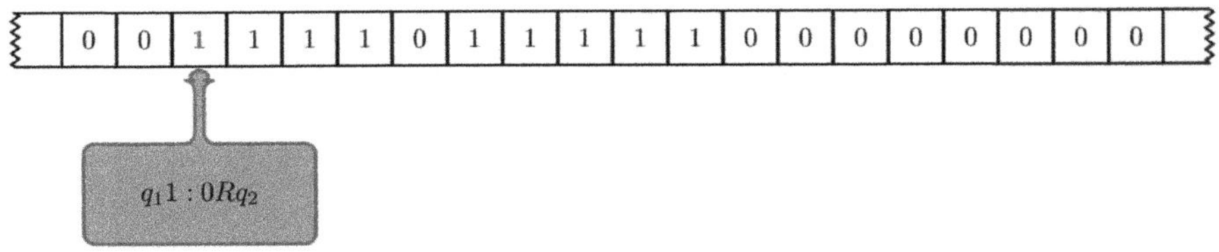

FIGURE 6.6. A schematic representation of a Turing machine, which is instate q_1 scanning the symbol 1 and should in the next step print some symbol, move one square to the right and move to state q_i. Image courtesy of Liesbeth De Mol.

In the 1940s, von Neumann and Turing used this and other similar models as a way to reflect on the possibilities and limitations of modern computers. However, the Turing machine die *not* serve as a steppingstone toward the development of the first computers in any essential sense. Later, however, theoretically minded computer scientists saw it as a foundation of the modern computer by projecting backward into history. This misunderstanding partially originated in the field of automata studies developed in the late 1950s which aimed, as did others, to develop formal models for computing and programming. That tradition has continued until today within the domain of theoretical computer science.

It was also within the setting of automata studies that the focus shifted to the efficient use of time and space for algorithmic methods that solve a specific problem, in particular by determining upper and lower bounds for algorithms. This practical concern soon led to the field of computational complexity theory, which classifies different types of computational problems according to their so-called space and time complexity. In this setting, the Turing machine has retained its importance as one of the most basic models used. One of the standard categories in computational complexity theory comprises those algorithms that can solve a problem in polynomial time (the class P). This means that the running time for the algorithm's program is bounded by a polynomial expression based on the input data. Closely related to this is the class NP, containing those algorithms which can verify in polynomial time whether a proposed solution to a problem is, in fact, correct. One might imagine that a problem in the NP category might not be solvable in polynomial time, but this remains a famous open question, the so-called P = NP problem, which asks whether these two classes are identical or not. Its present-day prestige can be gauged by the fact that the Clay Mathematics Institute chose it as one of the seven Millennium Prize problems. On the other hand, its difficulty can be judged from the circumstance that no one has yet come to collect the $1 million prize for solving it. This particular choice reflects part of a more general trend, namely, the rising prestige of (theoretical) computer science; today (pure) mathematics no longer stands alone as the (only) queen of the sciences.

While automata studies and computational models are concerned with the theoretical aspects of computation, including its limitations and possibilities, formal logic and mathematics have played a role in the context of computer programs and software design (Priestley 2011). Already during the early stages of their development, experts recognized that programming high-speed computers involved a number of daunting challenges. In particular, they understood that programming a machine using a low-level machine code would present extreme difficulties. This problem became even more acute once the commercialization of computers began to take off in the late 1940s and early 1950s. As the larger world of big business and industry became interested in exploiting the capabilities of these machines, it became obvious that simpler ways were needed to program them. Initially, in the 1950s, that demand led to the first programming languages, FORTRAN and LISP. Since most people involved with programming had a background in mathematics, these languages employed a good deal of mathematical and logical notation.

An important breakthrough came with the development of the ALGOL language (short for ALGOrithmic Language), which was influential in spawning later languages. ALGOL emerged from an international effort aimed at creating a *universal programming language,* a goal intended to counter the trend toward a proliferation of programming languages at the time.[12] Yet, ALGOL also served another purpose, in that it helped to give programming a mathematical identity (see the section following). In the 1960s, one of the leading magazines in computer science, *Communications of the ACM*, introduced a new section called Algorithms. Readers could here find "approved algorithms," expressed in the ALGOL language (an ALGOL program is not, of course, an algorithm). Published on a regular basis, these algorithms had to meet a specific standard to receive the magazine's seal of approval.

As more and more people began using computers, the need for such reviewed algorithms became increasingly apparent owing to a familiar problem. Many computer programs contained errors—a basic problem even today when our lives depend more and more on software engineering. A dramatic case involved the Boeing 737 Max planes, which partly because of a faulty software system malfunctioned, resulting in crashes that killed more than 300 people. In an effort to alleviate errors and remove bugs in large and complex software systems, programmers turned to so-called formal methods, i.e., the use of formal logic to verify the correctness of programs. Not everyone, however, found this approach convincing or even feasible. Dave Parnas, a software engineer, leveled harsh public criticism in the 1980s against the Reagan administration's Strategic Defense Initiative (SDI or "Starwars"), arguing that it was impossible to build a software system that could reliably counter a nuclear attack. Part of his argument was that formal verification would never be feasible (Clinton's administration abandoned SDI in the 1990s). Today, formal verification is one of two basic approaches used in safety-critical systems, such as self-driving cars.

Ironically, even though the popularity of correctness proofs waned in the 1980s, a renewed interest in formal verification arose through the development of so-called proof assistants, like COQ and other programming systems. These have historical roots in mathematical logic, including work from the 1950s on automated proofs of theorem. It was also in this same context that mathematicians like Thomas Hales entered the scene, since this kind of mathematical technology can provide formal proofs for theorems, like the sphere-packing problem (see previous section).

CHANGING ROLES AND DISCIPLINARY IDENTITIES

We began this chapter with comments about the lecture Kenneth O. May delivered in 1976 at a conference in Los Alamos on the history of computers. Speaking to an audience consisting of computer scientists, programmers and engineers. May sought to bring the history of computing into the history of mathematics, thereby confirming a popular view amongst academic computer scientists at the time that their field is about mathematics. Today, however, the history of computing appears to be very distinct from the history of mathematics—as reflected by the fact that very few historians really engage with that connection. Why is that the case? At least part of the answer seems anchored in changed relations between mathematics and the field of computing.

As we noted above, once it became clear that computation and the computer gave rise to a set of interesting new questions, leading practitioners began to search for an orientation for this emerging field of research. In that context, many looked toward mathematics as a foundation for the field, and several academic computer scientists even understood computing as a subfield of mathematics. In fact, as is clear from the introduction to the book that followed the Los Alamos conference, there was at the time a belief that theory was the captain and application the soldier:

> Few, except these giants, realized in the 1930s that the formalism of mathematical logic, considered by many a sterile exercise for philosophers and for mathematicians in search of a field, was instead the magic key to programming languages as well as to computer design. [. . .] Once more, we find a confirmation of the sentence [. . .] "Theory is the captain, and application the soldier."[13]

At the time of the conference in 1976, computing did not have the role it has today in science and society. Ever since that time, computing technologies have steadily expanded to the point where they have become ubiquitous. We rely heavily on them in our cars, in our phones, in our social lives, in our work places, in our healthcare and justice systems. More and more, however, they hide inside other technologies. In fact, very few people are aware of the extent to which their lives depend on computer programs and their infrastructure and, thus ultimately, also on mathematics. Yet, more and more, personal, collective and political decisions have come to depend on these same technologies.

In this new setting, mathematics still has a key role to play, for instance by improving numerical methods within computing. This remains a major field of research, for example through advances in optimization algorithms for machine learning, though here its basic role is subservient to computing. At the same time, computing itself as a discipline and profession has matured and is steadily taking over much of the traditional role of mathematics as a tool for other sciences. Within the humanities, so-called "digital humanities" are growing in significance, just as in the exact sciences, simulation methods have become an established method to study and predict phenomena in biology, climate science, physics, etc. These trends have also affected mathematics itself: numerical analysis and computing within mathematics have climbed from their former lowbrow status to become well-respected and basic subfields. One can document this shift by citing statistics, namely, the growth in numbers of publications classified as belonging to numerical analysis or computer science. As an indication, in 2020, out of the 103,488 publications in pure and applied mathematics, fully 17,363 fell under the rubric of either computer science or numerical analysis, which represents around 17 percent of the total.[14]

Returning to May's perspective, one can only conclude that "mathematical technology," which for centuries was not a consciously recognized field of mathematics, has now fully matured. Indeed, it seems the roles of the science of mathematics and the technology of mathematics have been steadily reversing—a development one finds in many other fields since the 1980s. Perhaps, then, a new role will emerge for mathematics in the future. As society and science have become more and more driven and dominated by technologies, the techniques—mathematical and others—underlying them have become increasingly hidden away, as if closed up in a black box. This situation has only worsened since the advent of machine learning technologies, often promoted as some kind of magic, as oracles of the future. We cite here an editorial letter published in *Nature*, one of the most renowned scientific journals, in 2016: "The machine becomes an oracle; its pronouncements have to be believed [. . .] Intuitive machines will need more than trust: they will demand faith."[15] When even scientific discourse rises to the level of an almost religious belief in computing technologies, we may hope that mathematics may help to open up these black boxes in order to show us there is nothing magical about them. Just as basic mathematics education in the late eighteenth century was a fundamental part of a general alphabetization of society, so might educators in the future call upon mathematics to play a similar role by promoting a new form of literate numeracy.

CHAPTER SEVEN

Representing Mathematics

Mathematics in Twentieth-Century Literature

NINA ENGELHARDT

"Two plus two makes four." This simple sum has become a symbol for certainty, inevitability, and unalterable fact. It appears as an example of archetypal truth as early as the sixteenth century, and its "opposite"—"two plus two makes five"—similarly has a long history of designating the false and absurd. Literary texts in the twentieth century use the slogan mainly in two ways: to sketch mathematics as an ideal, rational system of truth, and to mark concern with its application to real, human matters. The probably best known instance of the slogan is George Orwell's *Nineteen Eighty-Four* (1948), a dystopian novel in which the protagonist Winston Smith attempts to hold on to this mathematical truth in a society that manipulates facts, memories, and perceptions to suit the purposes of its leaders. In this repressive system that almost annihilates individual liberty, Winston takes the equation as a symbol for the existence of truth that the totalitarian Party cannot touch: "*Freedom is the freedom to say that two plus two make four*" (Orwell 1990: 84, original italics), he states. His belief in the simple sum is the last conviction to go under torture, and Winston's ultimate acceptance of "2 + 2 = 5" signals the victory of the state and its unlimited power to pervert even the most certain truth of mathematics.

A comparison with Fyodor Dostoyevsky's novella *Notes from the Underground*, published about a century earlier in 1864, suggests how much Orwell's depiction of mathematics is influenced by the experience of twentieth-century totalitarianism, particularly Stalin's brutal elimination of opposition and manipulation of documents. Where Orwell presents math as the last bastion of truth and freedom in a society that perverts all facts, in the eyes of Dostoyevsky's narrator, the certitude that "two plus two makes four" is imprisoning: "what will have become of our wills when everything is graphs and arithmetic, and nothing valid but two and two make four? Two and two will make four without any will of mine!" (Dostoyevsky 2003: 39), he complains and emphasizes that "twice two is four is not life, gentlemen, but the beginning of death" (ibid.: 40). Dostoyevsky's text is an example of a broader tendency in late nineteenth- and early-twentieth-century literature and society to view rationality critically, as stifling emotion, individuality, and creativity and being opposed to often surprising and irrational human life. *Notes from the Underground* also clearly juxtaposes the constraining certainty of math with the unreliable narrator who recounts the story according to his subjective experience and with creative license, and thus establishes an opposition between "dead" mathematics and creative literature.

Many texts in the decades around 1900 contribute to the image of math as the rational, unalterable, and "dead" opposition to irrational, undetermined, emotional life and employ it in their critiques of industrialization and modernity with its increasing reliance on minute organization, science, and technology. For example, D. H. Lawrence draws on these associations when he describes a town with a "great, mathematical colliery on the other side" (Lawrence [1915] 2011: 321). The, on the face of it nonsensical, description—what is a "mathematical colliery"?—establishes math as part of a dehumanizing industrialization and modernization. Lawrence and Dostoyevsky employ math metaphorically, as an image of certain order and dehumanization, while other texts apply it directly to human concerns. The most sustained example of the latter tendency is Yevgeny Zamyatin's dystopian novel *We* (1921), which features a repressive society based on mathematical rules. In *We*, citizens, called "figures," have numbers as names, make decisions rationally to benefit the stability of the community, and live according to the law of the multiplication table. The first-person narrator D-503 explains:

> The multiplication table is wiser and more absolute than the ancient God. It never – repeat, never – makes a mistake. And there's nothing happier than figures that live according to the elegant and eternal laws of the multiplication table. No wavering, no wandering. Truth is one, and the true path is one. And that truth is two times two and that true path is four. And wouldn't it be absurd if these two happily, ideally multiplied twos started thinking about some kind of freedom, that is, about some mistake? (Zamyatin 1993: 65–6)

The juxtaposition of mathematical truth and individual freedom is the crucial dystopian aspect of this society: the text shows that the rigid, rational mathematics by which the society is organized, leaves no room for human factors such as individuality, unconscious drives, and emotions such as love. However, Zamyatin's *We* does not formulate a critique of mathematics itself but of its application to human matters. It vehemently denounces a society built on mathematical rules, yet also suggests that reducing math to the multiplication table is doing it a disservice. The protagonist D-503 discovers that irrational numbers and imaginary numbers are part of mathematics too and that it thus includes these major factors of human life and even "proves" their ineradicable reality that the totalitarian society aims to suppress. Similarly, Margaret Atwood's dystopian novel *The Handmaid's Tale* (1985) points to the inadequacy of applying mathematical rules to human affairs when the narrator sets out the impossibility of adding up individuals to a unified sum: "One and one and one and one doesn't equal four" but "just one and one and one and one" (Atwood 2010: 201, 195). Individuality and human life cannot be reduced to a number, exchanged, or lumped together in a sum, Atwood's novel suggests: "They cannot replace each other. Nick for Luke or Luke for Nick" (ibid.: 202). Like Zamyatin, Atwood does not blame mathematics as intrinsically destructive but objects to applying it to human life.

While mathematics plays only a modest role in some of the literary texts discussed so far, these examples show how it is employed as a shorthand for reason and twentieth-century concerns with revaluations of rationality and possibilities for individuality and freedom. Almost all texts to some degree acknowledge the common juxtaposition of mathematics and, on the other side, life, literature, and creative freedom, even if only to take it as a basis from which to ultimately denounce the contrast and the reductive image of math. The remainder of this chapter is dedicated to exploring the representation of mathematics in works with literary ambition that have neither a mainly didactic nor a purely entertaining intention, and the chapter will discuss only in passing topics such as mathematics in popular culture, scientists' biographies, or hard science fiction. The focus lies instead on literary texts that represent math in more complex terms, revealing surprising aspects of uncertainty, freedom, and creativity in it; suggesting similarities between math and literature; and showing that applying mathematics to life can also yield fascinating insights into its mysteries.

LANGUAGE AND TEXT: MATHEMATICS AS A MODERN MEANS OF EXPRESSION

In Orwell's *Nineteen Eighty-Four*, Winston Smith, having capitulated after months of torture, acknowledges his defeat by writing "*TWO AND TWO MAKE FIVE*" (Orwell 1990: 290) on a piece of paper. Yet, he secretly prepares to keep his oppositional feelings hidden until moments before his death and then release them and die as a free man. After another round of torture wipes out all remaining resistance, he absentmindedly scrawls "2 + 2 = 5" (ibid.: 303) into the dust. The involuntary act confirms that the Party has now taken possession of his unconscious, and the change from writing words—"two" and "five"—to noting down numbers as symbols—"2" and "5"—further designates that he has internalized his status as a mechanized, depersonalized cog in the machine of the totalitarian system. As this example shows, numbers stand out against words in writing and have different connotations, and the symbolic language of mathematics affords possibilities of expression that differ from natural language. While employing words integrates math into the world of the literary text, numbers draw attention to the "strangeness" of mathematical symbols as well as to the typography and materiality of the text as a whole. If stories can be understood as windows into a different, fictional world, the appearance of non-literary characters draws attention to the "window glass," to the text itself: a mathematical equation highlights that we look at a printed text. This effect made mathematical writing attractive to early twentieth-century poets, and mathematics features in poems and essays by writers such as T. S. Eliot and Filippo Tommaso Marinetti.

Modernist writers deliberately searched for ways to make literature and language "new," and for a number of authors, mathematics provided a means of doing so. Given the dramatic changes at the beginning of the twentieth century—immense technological and scientific developments, shifting gender and class relations, the First World War that upset the notion that the world was on a path of progress and betterment—modernist writers felt that traditional, realist literature was no longer adequate to represent the world and that innovative forms of literary expression had to be found.[1] In "Triumphal March" (1931), a poem by T. S. Eliot that responds to the First World War, the first half of the second stanza is taken up by a list of weapons and military equipment. Each of the ten short lines follows the same structure: "5,800,000 rifles and carbines / 102,000 machine guns, . . ." (Eliot 2015: 131, lines 14–15). Tens of thousands of mortars, airplanes, ammunition and field wagons, field kitchens, and field bakeries are listed, interrupted only by the assertion that the speaker cannot guess the enormous number of projectiles, mines, and fuses. The interjection as the only line that does not include a figure draws all the more attention to the fact that the numbers stand out in the poem and thus add to expressing the extraordinary, upsetting scale of the war.

Yet, Eliot was also interested more generally in how mathematical language could work as a model for innovations in prose writing. He was particularly impressed by Bertrand Russell and Alfred North Whitehead's *Principia Mathematica* (1910–13) and celebrated its contribution to making English "a language in which it is possible to think clearly and exactly on any subject. The *Principia Mathematica* are perhaps a greater contribution to our language than they are to mathematics" (Eliot 1927: 291). Eliot praised math as objective, precise, and impersonal and thus as a model for a new, modern way of writing. Similarly, the Futurist poet Filippo Tommaso Marinetti turned to mathematics as a way of breaking with conventional poetry and to "liberate words" (Marinetti [1912] 2009: 119). Next to abandoning syntax, infinitives, and adjectives, Marinetti called for replacing punctuation with mathematical symbols to indicate movement and direction: "mathematical signs will be used: + - x : = > <" (ibid.: 120). Here and elsewhere in early twentieth-century writing, math is presented as exemplary of abstraction, precision, and economy of formulation, and thus as an emblem for new literary ways of expression.

Marinetti explains his use of mathematics symbols:

> I create true theorems or lyrical equations, introducing numbers which I've intuitively chosen and placed within the very center of a word; with a certain quantity of + - x =; I can give the thicknesses, the mass, the volumes of things which words otherwise have to express. The arrangement + - + - + + x, for example, serves to render the changes and accelerations in speed of an automobile. (Marinetti [1914] 2009: 180)

His poster poem "montagne + vallate + strade x Joffre" (Mountains + Valleys + Roads x Joffre) (1915), later called "Après la Marne, Joffre visita le front en auto" (After the Marne, Joffre Visited the Front by Car), uses mathematical signs to illustrate Marshal Joseph Joffre's visit of the battlefield of the Marne in September 1914.

The composition of various typefaces and sizes of letters, words, numbers, and the mathematical symbols +, -, x, =, and a radical sign "√" visually express the battle and Joffre's journey. Among the "ta-tatatatatata ta" that onomatopoetically renders machine guns, the letter "M" that visually suggests mountains and the letter "S" or integral sign "∫" for the river Marne, "+"-signs emphasize the predominantly positive values of the words "BELLE" and "Mon Amiiiii", while the diagonal line of " + + - + - x - + x x x -" might render the changing speed of Joffre's car, and the lists of numbers point to the tactical calculations of material and soldiers' lives. The mathematical symbols, together with the "words in freedom" that are liberated from conventions of typography and sense, narrate the story of Joffre's journey in a new, visual form. Such use of mathematical signs in Futurist poetry disrupts traditional ways of reading, signaling the need to find new ways of looking at and representing a world in turmoil. In this way, Futurist poetry presents math as differing from everyday discourse in its precision and conciseness and as an exemplary means of modern expression.

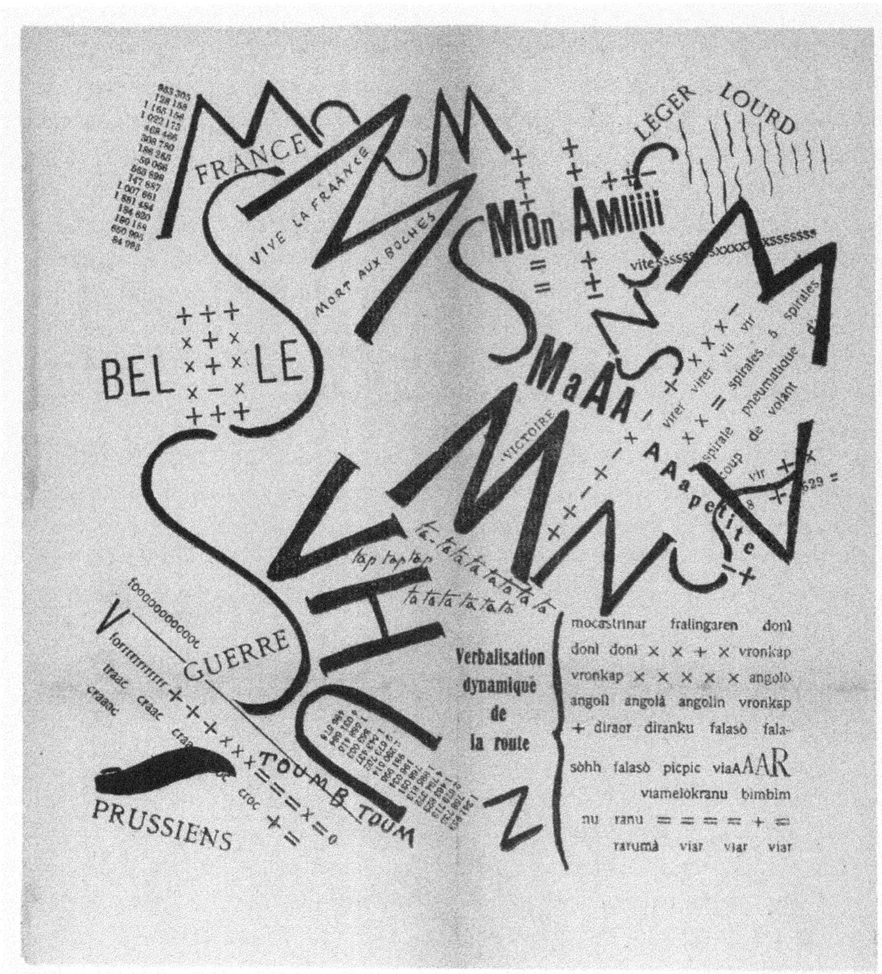

FIGURE 7.1. Filippo Tommasso Marinetti, "Après la Marne, Joffre visita le front en auto" ["montagne + vallate + strade x Joffre"] (1915). Photo by Fine Art Images / Heritage Images / Getty Images.

Using mathematical notation in a literary text and thus disrupting the usual page layout and reading flow sometimes have more mundane reasons and effects. Thomas Pynchon's novel *Gravity's Rainbow* (1973) includes an equation introduced as "hilarious graffiti of visiting mathematicians," a joke that relies on the blending of everyday and mathematical language:

$$\int \frac{1}{(\text{cabin})}\, d(\text{cabin}) = \log \text{cabin} + c = \text{houseboat}$$ (Pynchon [1973] 2000: 534)

Readers have to verbalize the equation to appreciate the joke: a log cabin plus c (homophone of "sea") equals a house on the water or a houseboat. As this example demonstrates, literary texts do not always include mathematical notation to symbolize rationality or precision or to achieve an estrangement effect. They can also deflate its aura of difficulty and more closely connect mathematical and literary language.

BEYOND REALITY? NON-EUCLIDEAN GEOMETRIES AND THE FOURTH DIMENSION

A "qualitative imaginative futurist mathematics"—this is what Marinetti, in collaboration with the mathematician Marcello Puma and the physician Pino Masnata, aimed to achieve (Marinetti [1941] 2009: 298). In a short, hyperbolic manifesto, the trio claims that "Italian Futurism is renewing even mathematics," namely into an "antistatic antilogical antiphilosophical mathematics" (ibid.: 298) that is suitable to the dynamic, irrational, concrete modern world. The authors call for mathematicians to "appreciate new geometries and gravitational fields created by masses moving with sidereal velocity" (ibid.: 299), that is, to take account of "new" non-Euclidean geometries and relativity theory. They here identify a classical and a new, modern strand of math and interpret specific developments in the late nineteenth- and early twentieth centuries as particularly consequential. Other literary writers, too, allude to modern changes in mathematics itself and praise its departure from long-held beliefs.

The "new geometries" that Marinetti and his co-authors highlight emerged in the nineteenth century when the discovery of non-Euclidean geometries by János Bolyai and Nikolai Lobachevsky broke with the geometry thought valid for over two thousand years. Geometry literally means "measurement of the earth," from Ancient Greek: *gê-* "earth" and *metría* "measurement", and the formalized theory in the Greek mathematician Euclid's *Elements* had its origins in land measurements. Nineteenth-century mathematicians realized, however, that they could dispense with Euclid's so-called parallel postulate. That meant that they could design other geometries, non-Euclidean ones. These are as mathematically consistent and thus just as "correct" as Euclidean geometry, and this created a situation of uncertainty: if there are several valid geometries, which one is the "right" one? Mathematicians concluded from this and other discoveries that mathematical ideas should not be thought of as the true language of nature—as Galileo had once famously proclaimed—but can best be understood as a self-referential system that follows its own logic independently of physical reality. Literary writers saw in this an analogy to fiction, which similarly creates its own rules apart from the real world. For example, Zamyatin presents Lobachevsky's non-Euclidean geometry as leading the way to a new form of literary representation that does not try to "copy" the visible world realistically but develops free from former conventions.

In an essay from 1923, Zamyatin emphasizes the revolutionary potential of what he calls the "new mathematics":

> Two dead, dark stars collide with an inaudible, deafening crash and light a new star: this is revolution. [. . .] Lobachevsky cracks the walls of the millennia-old Euclidean world with a single book, opening a path to innumerable non-Euclidean spaces: this is revolution. (Zamyatin 1970: 107)

He goes on to argue that if modern math breaks with former certainties and reveals new spaces, so should literature. Instead of adhering to traditional realist conventions of trying to faithfully depict objective reality, modern literature should account for "new spaces" such as inner life, unconscious desires, and irrational drives:

> Science and art both project the world along certain coordinates. [. . .] All realistic forms are projections along the fixed, plane coordinates of Euclid's [two-dimensional] world. These coordinates do not exist in nature [which is three-dimensional]. Nor does the finite, fixed world; this world is a convention, an abstraction, an unreality. And therefore Realism [. . .] is unreal. Far closer to reality is projection along speeding, curved surfaces – as in the new mathematics. (Ibid.: 112)

In *We*, Zamyatin translates this agenda into fiction by developing from a rational, objective writing style to diction that is fragmented, disordered, and shifts suddenly. With this new form of expression and a focus on the characters' irrational and emotional lives, he replaces conventions of realism with a modernist style that takes account of the formerly disregarded non-rational inner spaces. In the novel, the protagonist D-503 compares his new writing style not to non-Euclidean geometries but to the unknown and imaginary in mathematics:

> Instead of the elegant and strict mathematical poem in honor of OneState [the totalitarian state D-503 lives in], it's turning out to be some kind of fantastic adventure novel. Oh, if only this really were just a novel instead of my actual life, filled with x's, $\sqrt{-1}$, and degradations. (Ibid.: 99)

Using the mathematical symbol x that often designates an unknown value and the imaginary number $\sqrt{-1}$ to metaphorically describe the unknown and imaginary aspects of life, *We* establishes math as a realm that goes beyond immediately perceivable reality and is reflected in a literature that similarly does so.

Visual artists were also inspired by the new spaces that nineteenth-century mathematics introduced. In particular, the fourth dimension fired the imagination, first as a further spatial dimension and, after the popularization of Albert Einstein's theory of relativity in the 1920s, as time in a four-dimensional space-time continuum. Unlike Marinetti's visual poster poems, Cubist painting does not employ mathematical symbols, but it evokes a sense of the fourth dimension in its style of representation. For example, in *Les Demoiselles d'Avignon* (The Young Ladies of Avignon, 1907) Pablo Picasso makes up the bodies of five ladies from geometrical shapes and reassembles the triangular, semi-circular, and polyangular slices so that viewers see the same face from different perspectives at the same time.

With its presentation of different views not normally seen simultaneously, Picasso's picture translates a viewpoint from the fourth dimension onto a two-dimensional canvas. Showing a face both in profile and as a frontal view

FIGURE 7.2. Pablo Picasso, *Les Demoiselles d'Avignon*. Photo by STAN HONDA / AFP via Getty Images.

radically departs from established artistic conventions, and this rendering of perspectives and knowledge that are simultaneously available only in the fourth dimension, signified a liberation of painting from the rules of realistic representation. Cubist art thus paints an image of mathematics as including dimensions beyond experienced three-dimensional space and as unlocking new possibilities of perception that lie beyond visual reality. Departing from visual reality and traditional viewpoints more broadly inspired innovation in modern art movements, and literary writers also aimed at representing the multiple dimensions described by mathematics, for example by experimenting with simultaneity and multiple narrative perspectives. Literature also responded to the suggestion of an invisible but no less real dimension beyond sensual experience by invoking the fourth dimension to account for phenomena beyond material existence, such as consciousness or the spiritual habitat of beings such as ghosts or the living dead.[2] Overall, mathematics here appears as a realm that reveals aspects beyond physical reality and inspires leaving behind ties to the real world and exploring what is beyond the scope of material reality.

THE "FOUNDATIONAL CRISIS" AND THE IMAGINARY WORLDS OF MATHEMATICS

The symbol of undeniable truth, 2 + 2 = 4, became far less self-evident in view of developments in the nineteenth century that, for example, questioned the seemingly obvious but ultimately challenged validity of Euclidean geometry. To prevent such surprises, mathematicians investigated the foundations of their field and aimed to set it on a secure basis without reference to the physical world. In their *Principia Mathematica* that so impressed T. S. Eliot, Bertrand Russell and Alfred North Whitehead set out to prove that 1 + 1 = 2. Since it takes them 362 pages to arrive at this proof, this demonstrates that simple sums might appear to be self-evident but that proving their validity is far more difficult. The efforts of Russell, Whitehead, and other mathematicians to provide secure foundations for their field revealed surprising complications and paradoxes. Indeed, these difficulties suggested—and in 1931 this was even proved with Gödel's incompleteness theorem—that it is impossible to define a complete and consistent set of axioms for the foundations of mathematics. Many mathematicians remained unfazed by these developments: 1 + 1 = 2 gave reliable results even before Russell and Whitehead's proof, and this suggests that an unproven or even unprovable state does not necessarily have practical implications. However, the situation of working in a field whose basic grounds remained uncertain did disconcert some professionals, and the resulting sense of anxiety also traveled outside the sphere of practitioners. The implications of this so-called "foundational crisis of mathematics" are profound: if math is set on unproven foundations, then everything based on mathematical calculations, including science and technology, is similarly built on insecure grounds. Moreover, questioning the truth of the most rational form of thinking fueled doubt that it is possible to attain certain knowledge at all. As David Hilbert, one of the most prominent mathematicians of the early twentieth century, said in 1925: "If mathematical thinking is defective, where are we to find truth and certitude?"[3]

While only some mathematicians expressed concerns of this kind, the notion of a foundational crisis of mathematics fit into a list of areas that were seen to suffer from a loss of certainty in the decades around 1900: religion faced increasing questioning of God's existence, confidence in the progressive development of the world received a severe blow with the atrocities of the First World War, and literary writers as well as non-professionals lost trust that language stably refers to the world and is a reliable means of communication. Yet, while works by literary writers such as Robert Musil, Hermann Broch, and Thomas Pynchon present the mathematical crisis as a blow to visions of certain knowledge and truth, they also identify in it possibilities for creative change and freedom. Robert Musil's "The Mathematical Man" (1913), an essay in the voice of a fictional narrator, begins by pointing out that life essentially relies on mathematics, as it provides the basis of science and technology and thus for machinery necessary in everyday life, from ovens to bake bread to architects' calculations to build homes. The questioning of mathematical certainty in the "foundational crisis" then threatens all of life, so the narrator claims:

> Suddenly, after everything had been brought into the most beautiful kind of existence, the mathematicians [. . .] came upon something wrong in the fundamentals of the whole thing that absolutely could not be put right. They actually looked all the way to the bottom and found that the whole building was standing in midair. (Musil 1990: 42)

Everyday life, which relies on mathematics for its buildings and meal preparation, similarly finds itself lacking a foundation. Yet, so the narrator suggests, if mathematicians reveal the precarious situation we live in, they also demonstrate how to deal with it: they ward off panic about the missing foundations by pragmatically focusing on the fact that the system still functions and provides real results, and they just continue with their work. The narrator proposes a similar path for the literature of the early twentieth century, which he characterizes as having abandoned reason while focusing entirely on feeling. Musil's own novel, *The Man without Qualities* (1930/32), attempts to hold fast to the benefits of reason, question its foundations, and trust in reality although it might hover in midair. Many literary works of the twentieth century share this concern with a renegotiation of reason and its opposite, variously

FIGURE 7.3. Mason-Dixon Line Marker for the border between Maryland and Pennsylvania. Photo by Charles Phelps Cushing / ClassicStock / Getty Images.

understood as mysticism, emotion, or imagination. The majority of these texts do not refer to mathematics directly but those that do often locate it in a broader modern period of crisis and debates on reason and rationality. While the "Age of Reason," as the eighteenth-century Enlightenment is often called, celebrated the value of rationality, in the twentieth century, the hopes placed in the rational faculties of human beings received a significant dent and even reversal.

In Pynchon's novel *Mason & Dixon* (1997) math features at once as a method and symbol of Enlightenment rationality and its destructive power and as a "postmodern" realm that does not provide objective knowledge. In the latter sense, it exists in diverse, socially and culturally constructed versions, and reveals different worlds. The novel is set in the eighteenth century, a period of great trust in reason that engaged in a mathematization of the sciences, which proved immensely successful for consolidating mathematics as the pinnacle of reason. *Mason & Dixon* views this period from the vantage point of the late twentieth century and critically illuminates the downsides of the Enlightenment, including its reductive effects on mathematics itself. Although based on the historical figures Charles Mason and Jeremiah Dixon, it overtly changes and fictionalizes events, complements facts with blatant anachronisms, and introduces undeniably fictional entities such as ghosts, talking mechanical ducks, and flying mathematicians. The main part of the novel presents the two principal characters in their task of surveying the Mason-Dixon line that separates the states of Maryland and Pennsylvania. Many characters see Mason and Dixon as bringing rational order to the wilderness of the "New World" of America: "pure Space waits the Surveyor,– no previous Lines, no fences, no streets to contain polygony however extravagant," a character comments (Pynchon 1997: 586). The two surveyors calculate where the boundary line has to go, draw it on a map, and this comes to mark a real separation between the states while disregarding organically grown local lines of demarcation such as the Great Indian Warpath. In this sense, calculations on paper forcefully bring into being the new reality of America.

In Pynchon's novel, there are different "mathematicses" used to measure, calculate, and thus "create" America: rational, Enlightenment mathematics and, on the other hand, a math that is mysterious, medieval, and close to magic. This latter version is mostly related to the Chinese Cult of Feng Shui. As Mason and Dixon travel to America—the "New World" that is still being explored in the eighteenth century—they encounter both the 360-degree-circle they are used to and Feng Shui circles that sport 365.25 degrees:

> "Mr. Dixon," declares the Jesuit, "at present, owing to the pernicious Cult of *Feng Shui*, you would find it a Surveyor's Bad Dream,– nowhere may a Geometer encounter an honest 360-Degree Circle,– rather, incomprehensibly

and perversely, in willful denial of God's Disposition of Time and Space, preferring 365 and a Quarter."

"That being the number of Days in a year, what Human Surveyor, down here upon the Earth, would reject thah',– each Day a single, perfect Chinese degree,– were 360 not vastly more convenient, of course, to figure with?" (Pynchon 1997: 229)

As Mason and Dixon put their net of unnatural but convenient Enlightenment mathematics over America, the Feng Shui circles disappear, both from mathematics and from nature. The rational, Enlightenment math that the surveyors force onto nature—and they do so literally, hacking down trees and

FIGURE 7.4. A Chinese geomancer's compass.Photo by SS PL / Getty Images.

leaving the land with "geometrick Scars" (Pynchon 1997: 257)—determines the landscape and makes circles with 365.25 degrees disappear. In other words, Pynchon's novel presents mathematics as having effects on reality: math does not simply convey knowledge of the world, but it shapes and creates it.

Differently shaped circles might seem an amusing oddity in fiction, but with another example of mathematics forming reality—this time with a basis in historical fact—Pynchon's novel clarifies what is at stake: it uses the Vortex to illustrate how mathematization impacts on reality. René Descartes' vortex theory offered an explanation of planetary motion and gravitation—essentially, it is a theory of the universe. Descartes believed that the universe is permeated by an ether, that is, a material substance that, according to ancient thought, fills the upper regions of space. Descartes further held that the ether is organized in circling bands of particles he called vortices.

By the mid-eighteenth century in which Pynchon's novel is set, Cartesian vortex theory had become superseded by the theories of Isaac Newton, a main reason being that, whereas Descartes used verbal formulations to explain phenomena, Newton developed his theories in mathematical form. Newton's work was a driving factor in the transformation of mechanical explanation, and the very title of his masterpiece—*Philosophiae Naturalis Principia Mathematica* (Mathematical Principles of Natural Philosophy) (1687)—announces his aim of using mathematics to interpret physical phenomena. His mathematically formulated "law of universal gravitation" proved immensely useful even though Newton was unable to give a mechanical explanation of gravity: "the cause of gravity is what I do not pretend to know," he wrote.[4] Pynchon's earlier novel, *Gravity's Rainbow* (1973), explores questions and implications around this phenomenon, and *Against the Day* (2006) is concerned with the abandonment of the ether in the late nineteenth century. *Mason & Dixon* includes in its fictional exploration the displacement of vortex theory by Newton's mathematized physics.

In Pynchon's novel, Dixon's worship of Newton derives from his teacher William Emerson, a character whose historical counterpart is the British mathematician of the same name. As a follower of Newton, Emerson opposes Vorticism, enjoying currents and vortices in a river but objecting to them in cosmology:

> He loves Vortices, may stare at 'em for hours, if he's the Time, so far as they remain in the River,– yet, once upon Paper, he hates them, hates the misuse,– and therefore hates Euler, for example, at least as much as he reveres Newton. (Pynchon 1997: 220)

Dixon's education by Emerson explains why he is not among those characters in Pynchon's novel who see "the Vortex." In contrast, Mason believes in and

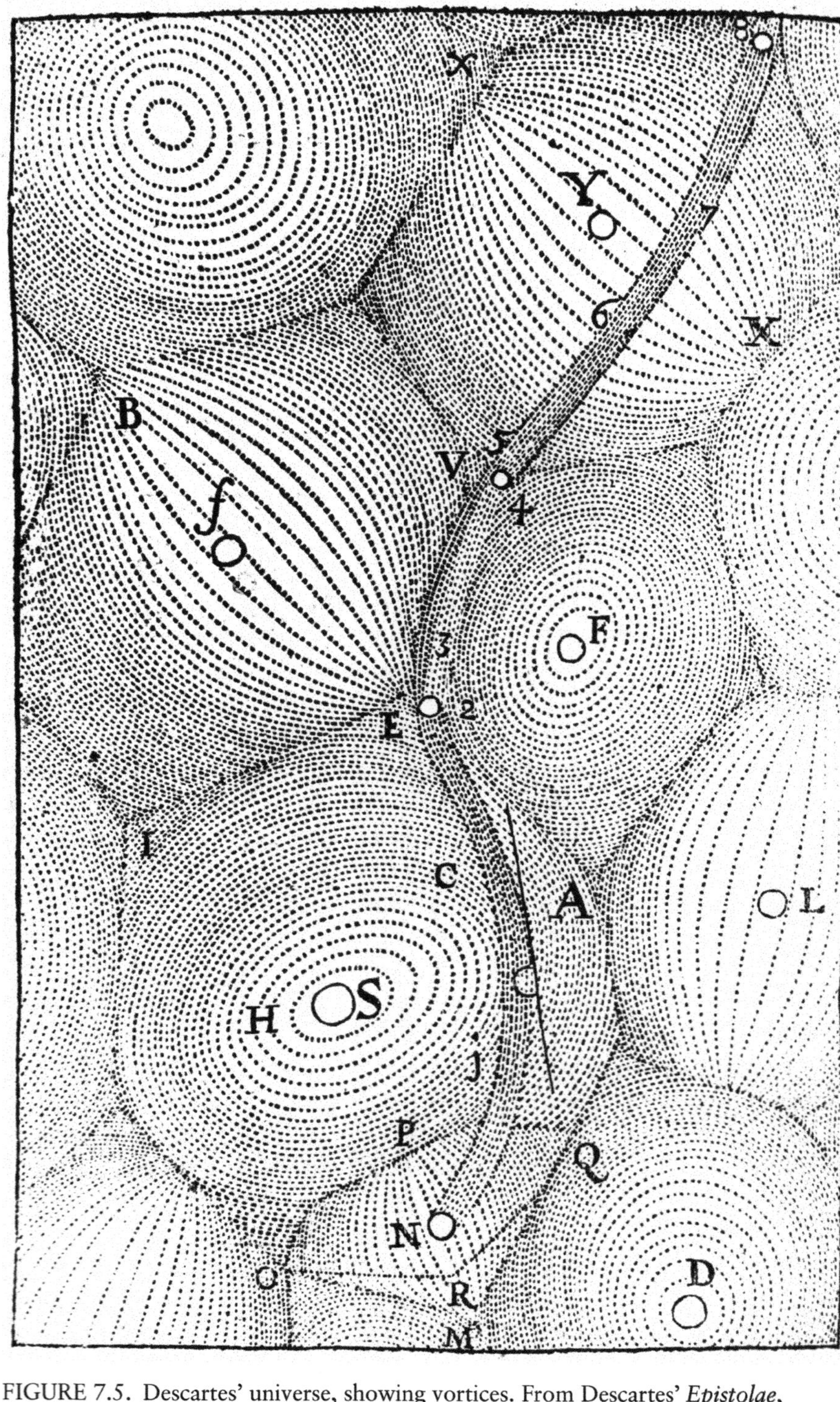

FIGURE 7.5. Descartes' universe, showing vortices. From Descartes' *Epistolae*, Amsterdam 1688. Photo by Universal History Archive/Getty Images.

enters the Vortex, doing so during a calendar reform. As part of Enlightenment standardization, Britain changed from the Julian calendar to the Gregorian calendar in 1752. This meant that September 2 was immediately followed by September 14. In the novel as well as in historical reality, some of those affected by the reform complain that they lose eleven days of their lives, but the fictional Mason does not advance together with the calendar. He lives through the eleven "missing days," passing the time that does not fit the rationalized calendar in the Vortex, a space that similarly contradicts Enlightenment standards.

Mason recounts having entered into "a slowly rotating Loop, or if you like, Vortex, of eleven days, tangent to the Linear Path of what we imagine as Ordinary Time, but excluded from it, and repeating itself,– without end," and Dixon speculates that the rotating Vortex creates gravitation (Pynchon 1997: 555). Functioning as an explanation of gravitation, the Vortex in *Mason & Dixon* shares characteristics with the Cartesian vortex and exemplifies the fate of explanatory concepts that were being displaced by Newton's mathematized physics. In the Vortex, Mason finds a counter-reality that is precisely opposed to Enlightenment explanations of the world and accommodates "all that Reason would deny" (ibid.: 559) and threatens to disappear in the increasingly rationalized period. In historical reality, 1752, the year of the calendar reform, also marks a last defense of vortex theory against Newtonian science with the publication of *Théorie des tourbillons cartésiens* (Cartesian Vortex Theory) by Bernard le Bovier de Fontenelle. The preface to Fontenelle's anonymously published book seeks to defend vortex theory by attacking Newton's mathematical explanations and argues that mathematics is a form of knowledge, and indeed of being, that has no relation to the physical world:

> There is, so to speak, two very different worlds; one mathematical, the other physical. The mathematical [. . .] only exists in the ideas of the geometer: he supposes the infinitely small, the point without dimension, lines without width [. . .]; as well as vacuum and gravitation. [. . .] But nothing of this can be found exactly in nature [. . .] and it is a strange illusion to abuse of the abstractions in transporting them to the physical world as if they were real beings. (Anon. in Fontenelle 1752: x–xii, my translation[5])

While the preface contrasts an ideal mathematical and a concrete physical world, in *Mason & Dixon* there are two material worlds, one supported by Newton's mathematized science, the other a pre-Enlightenment world close to extinction and only existing for staunch Vorticists such as Mason. In both cases however, mathematics is presented as threatening a reality.

In 1996, when *Mason & Dixon* was published, it emerged in the climate of the so-called "Science Wars"—a fierce debate between scientific realists who defended the objectivity of science, and postmodern critics who questioned its objectivity by pointing to the role of social and cultural factors. The constructivist notion that science does not objectively explain natural facts but contributes to creating these, was widely discussed in the 1980s and 1990s, both approvingly and in shades running from unfavorable to appalled. The idea that mathematical approaches to nature lead to the *material* disappearance of that which is disproved or has otherwise fallen out of favor, was radical even for many constructivists. Yet, a brief discussion of a paper by historian of science Yves Gingras shows that, as exuberantly fictional as Pynchon's depiction of a material Vortex at the brink of disappearance might seem, it participated in a wider discussion about the truth and reality of mathematics and science. In his paper, "What Did Mathematics Do to Physics?" published in 2001, Gingras argues that mathematization had consequences firstly, for what counts as a proper scientific explanation, secondly, for assessing who could do science (namely only those proficient in math), and thirdly, that it led to the vanishing of substances, for example the Cartesian vortex. While the first two claims were relatively uncontroversial, many objected to Gingras' suggestion that "by its ever greater abstract treatment of phenomena, mathematization led to the vanishing of substances. Not only Cartesian vortices but also the luminiferous ether were dissolved in the acid of mathematics."[6] In other words, Gingras argues that what cannot be put into mathematical form ceases to exist. Irrespective of the correctness or falsity of the notion that discredited scientific concepts involve material disappearance, Gingras' views show that Pynchon's novel participates in a period of debate about the truth of mathematics, its construction of reality, and its relation to the physical world.

Mason & Dixon paints a picture of mathematics as a tool of Enlightenment rationality that simplifies complex reality by determining one definitive state and eliminating contradictory realities such as the Vortex: it helps "changing all from subjunctive to declarative, reducing Possibilities to Simplicities" (Pynchon 1997: 345). Eliminating the indeterminate and the possibilities inherent in it, Pynchon's novel portrays mathematics as a means to make disappear alternative views and worlds, particularly as the European Enlightenment meets and overpowers Chinese Feng Shui knowledge and Native American ways of life. The notion that mathematics makes alternative worlds disappear once again relates to the Cartesian theory of vortices. Descartes held that next to our solar system, there are indefinitely many other star systems and thus, worlds. While Descartes himself remained guarded regarding the possibility of earth-like and inhabited other worlds, many readers interpreted his work along these lines. At the time of Fontenelle's last defense of Cartesian vortex theory, he had already made a name with his writing on the possibility of multiple worlds in *Entretiens sur la pluralité des mondes* (Conversations on the Plurality of Worlds) (1686). As vortex theory stands for the possibility of a plurality of worlds, *Mason &*

Dixon suggests that connected with the disappearance of the Vortex is the vanishing of a plurality of worlds and a limiting concentration on the physical, visible, rationally explainable earthly environment.

Up to this point, the discussion of mathematics in Pynchon's *Mason & Dixon* contributes to an image of it as at the peak of rationality, led by Enlightenment endeavors to control nature, and opposed to indeterminacy, plurality, and openness. However, the novel juxtaposes this image of a limiting, rationalized math with a pre- and post-Enlightenment focus on the possibilities and plurality inherent in mathematics itself. Different characters embody different notions of math in the novel: Maskelyne, modeled on the historical mathematician and astronomer Nevil Maskelyne (1732–1811), embodies Enlightenment math and its hopes to find "some single gigantick Equation" of the universe (Pynchon 1997: 134). Emerson stands for the opposite when his research does not reveal greater order but enervating carelessness:

> He holds no quarrel with the Creator's sovereignty, but is repeatedly appall'd at the lapses of Attention, the flaws in Design, the squand'rings of life and energy, the failures to be reasonable, or to exercise common sense,– first appall'd, then angry. (Ibid.: 220)

Emerson is further marked as a pre-Enlightenment character when he practices mathematics as an occult craft close to magic that enables him and his pupils to fly. Pynchon here draws on historical accounts of Emerson as a quarrelsome, idiosyncratic man, whose neighbors saw him as a conjurer and annoyed him with appeals to recover their stolen goods or to predict the future. Emerson's habit of wearing his shirts back to front and his choice to write under the pseudonym "Philofluentimechanalgegeomastrolongo" certainly point to his eccentricity and suggest him as a perfect historical model of a "mathemagician"—part rational mathematician, part mysterious, pre-Enlightenment magician. The juxtaposition of two long-standing images of the mathematician—one the incarnation of reason and order, the other related to magic and idiosyncrasy bordering on madness—continues in a number of contemporary texts and films, such as *Proof* (2005), *The Imitation Game* (2014), and *The Man Who Knew Infinity* (2016).

In *Mason & Dixon*, mathematics also appears more open and pluralist than readers schooled in the Enlightenment tradition might expect, as when it includes circles with 365.25 degrees and the surveyors discover that, despite utmost precision, their mathematically measured lines leave a space that belongs neither to Maryland nor to Pennsylvania. The narrator explains:

> Geomancer, beware,– if thy Gaze but turn Eastward by an Eyelash's Diameter, though must view the notorious Wedge,– resulting from the failure of the

> Tangent Point to be exactly at this corner of Maryland, but rather some five miles south, creating a semi-cusp or Thorn of that Length, and doubtful ownership,– not so much claim'd by any one Province, as priz'd for its Ambiguity,– occupied by all whose Wish, hardly uncommon in this Era of fluid Identity, is not to reside anywhere. (Pynchon 1997: 469)

The importance of this mathematically created refuge becomes clear when the narrator explains that "to be born and rear'd in the Wedge is to occupy a singular location in an emerging moral Geometry" (ibid.: 323). The emerging moral division between the slave state Maryland and the free state Pennsylvania heavily foreshadows the American Civil War (1861-1865), which takes place about a hundred years after the land survey.

Moreover, the novel ends with an image of what the preface to Fontenelle's defense of vortex theory describes as treating mathematical abstractions "as if they were real beings." As Mason nears the end of his life, he no longer forces a mathematical web over nature but stays at home and discovers beings in math: "The beings who visited had names, and Titles, and signs of Recognition. Often they would approach through Number, Logarithms, the manipulation of Numbers and Letters, emerging as it were from among the symbols. . . ." (Pynchon 1997: 768). Pynchon's novel thus closes with a view of mathematics as a self-referential realm that is creative and can reveal other beings and worlds. In its pre- and post-Enlightenment manifestations, math does not lead to rationalization and simplification but instead uncovers possibilities that lie beyond the visible. Celebrating alternative spaces and worlds, including mathematical ones, Pynchon's novel calls to abandon the predominant image of math as a tool of the Enlightenment and to recover its surprising and creative aspects that do not illuminate the physical world but create worlds of its own. This image of what could be called a "postmodern" or "fictional" mathematics is less common in popular representations such as films. However, it also goes back to a long tradition of associating mathematics with art, and literary works explore this notion, particularly after the "foundational crisis" drew attention to uncertainty in mathematics and its relation to physical reality. The concept of imaginary numbers proves particularly fruitful for literary explorations of similarities between the imaginative worlds of literary fiction and mathematical realms that develop independently of reality. Several of the authors encountered in this chapter employ imaginary numbers as a metaphor: Musil in *The Confusions of Young Törless* (1906), Zamyatin in *We* (1921), and Pynchon in *Against the Day* (2006).

Imaginary numbers suggest a metaphorical connection between imaginative literature and mathematical entities that have no direct relation to physical reality. In his short story "Methodically Constructed" (1917), later integrated into his novel *The Guiltless* (1950), Hermann Broch goes further and compares the ways in which mathematical operations and literary fiction posit and thus "create" the conditions in which they operate. The short story uses mathematical expressions to "methodically construct" a plot about a math teacher who falls in love: "let the hero be localized in the middle class of a medium-sized provincial town [. . .] Let him be called Zacharias" (Broch 2000: 26–7). The story's setup echoes the introductory definitions in a mathematical operation—"let *a* be an even number" or "let C be a finite set of prime numbers"—and thus mirrors the process of mathematically "creating" being and setting. As Broch's short story, Pynchon's novel, and other twentieth-century literature suggest, mathematics and fiction can be seen to share characteristics of constructing beings and worlds that are internally consistent and do not directly relate to physical reality. In this way, literary texts present mathematics not only as a tool to rationalize the world into determinacy but also as a realm of creative freedom that can promote ideas of multiple worlds. This image of math can contribute to questioning the certainty and objectivity of knowledge threatened by the "foundational crisis" of mathematics and constructivist notions of scientific knowledge, and literary representations point to possibilities for recovering plurality in an age dominated by simplifying rationality.

MATHEMATICS AS METAPHOR AND MEANS TO REVEAL THE MYSTERIES OF LIFE: STOPPARD'S *ARCADIA*

Tom Stoppard's play *Arcadia* (1993) uses math—particularly chaos theory—as theme, metaphor, and formal device. This final section examines how these different levels work together to present math as an integral part of knowledge about human relationships and fascination with the mysteries of life. *Arcadia* introduces two time strands: one set in the years 1809 to 1812, the other in the present. In the nineteenth-century strand, 13-year-old Thomasina is a spirited girl and mathematical genius who offhandedly makes discoveries that in historical reality were only discovered years later. In the present-day setting, mathematician Valentine's explanations to historian Hannah help to ensure that the audience can grasp complicated mathematical ideas and their relation to life. Stoppard stresses that mathematics can be transferred to life, not in the sense of calculating human concerns but as a metaphor, that is, as an expression that equates two unrelated things because they share certain characteristics. In *Arcadia*, chaos theory with its elements of feedback loops, repetition, self-similarity, fractals, and self-organization works as a metaphor for human behavior. Not least due to the way it makes these mathematical intricacies relevant and fun, *Arcadia* was short-listed for "the best science book ever written" by the Royal Institution of Great Britain.

The beginning of the play introduces mathematics as a theme interrelated with several other main concepts—knowledge, literature, and sex. In a schoolroom of a country house, thirteen-year-old genius Thomasina is busy with a mathematics primer, but the first words to her poetry-reading tutor concern more practical matters: "Septimus, what is carnal embrace?" (Stoppard 2000: 2). In keeping with the spirit of the mathematics lesson, Septimus "proves" the relationship between this and sin: "when carnal embrace is sinful it is a sin of the flesh, QED" (ibid.: 2). Thomasina's demand to be taught "the true meaning of things" induces him to link mathematics and sexuality further, thus intimating that both have a "true meaning" that can be defined:

> Carnal embrace is sexual congress, which is the insertion of the male genital organ into the female genital organ for purposes of procreation and pleasure. Fermat's last theorem, by contrast, asserts that when x, y and z are the whole numbers each raised to power of n, the sum of the first two can never equal the third when n is greater than 2. (Ibid.: 4)

While the technical explanation is adequate and sufficient regarding Fermat's theorem, such a rational, simplifying definition of sex is less than satisfying. Not particularly mathematically inclined readers might think Thomasina's exclamation a fitting comment on math: "It is disgusting and incomprehensible" (ibid.: 4). While Thomasina is not unduly puzzled by the theorem, she still views sex as baffling, suggesting that logical questions unsolved for centuries can be far less enigmatic than the everyday mysteries of life. While this opening scene shows that separate spheres of knowledge such as mathematics and sexuality have specific conditions that call for different explanatory strategies, the rest of the play develops the ways in which math and sex can illuminate each other.

Despite her tutor's insufficient mathematical definition of sex, Thomasina tries to capture life mathematically. Dissatisfied with Euclidean geometry that reduces the intricate shapes of mountains to pyramids and trees to cones, she develops a "New Geometry of Irregular Forms"—her name for chaos mathematics—that describes more lifelike shapes: "If there is a curve like a bell, there must be an equation for one like a bluebell, and if a bluebell, why not a rose?" (ibid.: 49). Present-day character Valentine explains Thomasina's new geometry, which he understands to be akin to modern chaos theory:

> When your Thomasina was doing maths it had been the same maths for a couple of thousand years. Classical. And for a century after Thomasina. Then maths left the real world behind, just like modern art, really. Nature was classical, maths was suddenly Picassos. But now nature is having the last laugh. The freaky stuff is turning out to be the mathematics of the natural world. (Ibid.: 58–9)

Valentine here characterizes modern mathematics as a self-referential system, which, unlike Newton's classical work, does not aim to explain physical reality but "leaves the real world behind" just as Picasso's abstract painting ceases to realistically represent objects. Yet, mathematicians find that the theory developed without applications in mind nevertheless applies to nature in its irregular, chaotic features.

Chaos theory describes dynamical systems with high sensitivity regarding their initial conditions, as, for example, the weather where a small change in initial conditions—the proverbial flapping of a butterfly's wings—can make the difference between later seeing a gentle breeze or a hurricane. In *Arcadia*, Valentine explains Thomasina's discovery of the process of iterative calculation that describe these effects:

> The maths isn't difficult. It's what you did at school. You have some *x*-and-*y* equation. Any value for *x* gives you a value for *y*. So you put a dot where it's right for both *x* and *y*. Then you take the next value for *x* which gives you another value for *y*, and when you've done that a few times you join up the dots and that's your graph of whatever the equation is. [. . .] What she's doing is, every time she works out a value for *y*, she's using *that* as her next value for *x*. And so on. Like a feedback. She's feeding the solution back into the equation, and then solving it again. Iteration, you see. (Stoppard 2000: 57–8)

The graph to which Valentine refers, that is, the visual plotting of the results of the calculations, leads to intricate shapes, for example one resembling a leaf being "filled in" as more calculations produce more dots. The pattern that emerges from plotting such an iterative equation is a fractal and exhibits self-similarity: each "branch" of the leaf resembles the shape of the leaf itself, and each "branch" is, in turn, made up of smaller leaf-shapes that are, in turn, made up of even smaller leaf-shapes and so on to infinity.

As the same shape or pattern repeats itself on different levels, what at first looks random emerges as part of an elaborate order. In *Arcadia*, Valentine comments: "In an ocean of ashes, islands of order. Patterns making themselves out of nothing. I can't show you how deep it goes. Each picture is a detail of the previous one, blown up. [. . .] Pretty nice, eh?" (Stoppard 2000: 101)

In *Arcadia*, Thomasina does not have the time to calculate the many iterations needed for a shape to emerge, and when Septimus continues the calculations after her untimely death he similarly fails to see a pattern and, unable to prove that the universe has some kind of order, turns into a half-mad hermit. It takes Valentine's computer in the present-time strand to calculate the few million values for *y* that are necessary for an image to develop. Seeing the self-similar pattern with its "islands of order" emerge out of apparent disorder, and

FIGURE 7.6. "Barnsley Fern." Image courtesy of Nina Engelhardt.

wondering whether this could have given hope to despairing Septimus, Hannah asks:

Hannah "Do you mean the world is saved after all?"

Valentine "No, it's still doomed. But if this is how it started, perhaps it's how the next one will come."

Hannah "From good English algebra?" (Ibid.: 103)

Arcadia presents chaos mathematics as a way to probe the deepest questions about the universe and its order and meaning. It shows that classical Newtonian science, which implies a static, deterministic universe in which every future state could theoretically be calculated, does not account for a dynamic, irregular world. The modern, Picasso-like chaos theory more accurately describes dynamical systems and shows that a self-similar order evolves when viewing a large number of instances from the distance, an order that cannot be perceived on the basis of a small number of iterations and which does not allow for the prediction of individual incidents.

In its potential to reveal order in the universe, mathematics is meaningful for Thomasina, Septimus, and Valentine. The play also develops its initial contrast between mathematics and "carnal embrace" to suggest that math is relevant to human concerns as a metaphor for the unpredictable effects of individual behavior. Where scientists following Newton claimed that it should be theoretically possible to devise a world formula and thus predict everything to come—an idea formulated by Pierre-Simon de Laplace in 1814—the second law of thermodynamics states that not all operations are reversible. Once again, Valentine explains Thomasina's discovery of what she calls the "Improved Newtonian Universe" that takes account of non-reversible phenomena, must end in disorder, which is equivalent to growing cold:

> **Valentine** Listen – you know your tea's getting cold.
>
> **Hannah** I like it cold.
>
> **Valentine** (*ignoring that*) I'm telling you something. Your tea gets cold by itself, it doesn't get hot by itself. Do you think that's odd?
>
> **Hannah** No.
>
> **Valentine** Well, it is odd. Heat goes to cold. It's a one-way street. Your tea will end up at room temperature. What's happening to your tea is happening to everything everywhere. The sun and the stars. It'll take a while but we're all going to end up at room temperature." (Stoppard 2000: 104)

The discovery that taking into account heat necessitates a different view of the universe than Newton's is not a purely scientific one in *Arcadia*, but Valentine's sister Chloë arrives at it by observing human life: "The universe is deterministic all right, just like Newton said, I mean it's trying to be, but the only thing going wrong is people fancying people who aren't supposed to be in that part of the plan" (ibid.: 97). In contrast, Thomasina derives her "Improved Newtonian Universe" mathematically and sees its analogies in human life afterwards. Considering the promiscuous Mrs Chater as a human proof that Newton's laws of motion are incomplete since they do not account for the action of bodies in heat, she notes: "The Chater would overthrow the Newtonian system in a weekend" (ibid.: 112).

Introducing analogies between knowledge derived by mathematics and by observation of life, *Arcadia* establishes math as relevant to individuals, the understanding of the universe, and as far more able to enlighten the meaning of "carnal embrace" than its juxtaposition with Fermat's Theorem at the beginning of the play might have led to believe. Septimus' "proof" that "when carnal embrace is sinful it is a sin of the flesh, QED" (ibid.: 2) is tautological and thus meaningless, yet, mathematics does guide Thomasina's understanding as it provides her with a metaphor for the importance of "bodies in heat" and the unpredictably big consequences of details: Mrs Chater's affair with Septimus leads to her journey to the West Indies from where her husband sends a dwarf dahlia which, more than a 150 years later, disproves Bernard's theory that Mr Chater was killed in a duel with Lord Byron, leading him to exclaim: "Fucked by a dahlia!" (ibid.: 118). Using chaos mathematics with its characteristic sensitivity to details as a metaphor for human life, Stoppard's play suggests the unpredictability of the real world as individuals behave according to irrational desires and a small change can, like a butterfly's flapping wing, have enormous and unpredictable effects.

Arcadia discusses mathematics directly, both regarding details of mathematical operations and its application to and meaning in the world. Metaphorical uses illustrate analogies between math and human concerns, and while the play gives room to criticism of classical mathematics such as Euclidean geometry that does not adequately describe nature, the overall image is one of mathematics being relevant to and fully integrated into individuals' attempts to understand life. Even more however, math is integrated on the level of the play's structure and its value is thus communicated not only in the content but similarly through its form. For example, the process of iteration, feeding back the result of an equation into it, is mirrored by the structure of the play's two interrelated plotlines. In *Arcadia*, the past determines the present in two ways: it sets the conditions for the present, and it features as the aim of the present-day research when Valentine recovers Thomasina's mathematical discoveries and Hannah tries to determine whether the mad hermit in the records of the country house was Septimus. Valentine's explanation of mathematical iteration similarly applies to the structure of the play: "every time she [Thomasina] works out a value for y, she's using *that* as her next value for x. And so on. Like a feedback. She's feeding the solution back into the equation, and then solving it again" (ibid.: 57–8). On the plane of the play's plot, the first y is the past, which originated from an unknown "algorithm of life" and input. Taking the past as the new source and feeding it back into the algorithm of life, the present-day action ensues. In its turn it tries to recover the past and the algorithm of life, asking in particular whether its final result will imply meaningful order or a disordered, lukewarm universe. If, as Valentine shows Hannah, iterated algorithms form self-similar fractals, the play can be understood as a fractal: the present-day action incorporates the past, and with many parallelisms of characters, relationships, dialogues, and inquiries, the plot-strands exhibit self-similarity. *Arcadia* could thus be described as a fractal play, and the huge effects that minor details in the nineteenth-century plot have on characters in the present day deepen the analogy.

Arcadia exhibits a fractal structure, and as mathematical fractals are the result of multiple iterations of a chaos equation, the development of the plot spells out chaos theory. Chaos theory describes "deterministic chaos" since, contrary to the common perception of chaos as random, the development in a chaotic system is not arbitrary but just too complex to be predicted. Theatre-goers directly experience simultaneous determination and unpredictability when watching Stoppard's play: while the audience is very unlikely to predict the development of the plot, everything is predetermined as the text is already written—in this sense, the theatre performance of *Arcadia* follows the rules of deterministic chaos. Moreover, the stage visually illustrates the growing disorder of the universe that will end up at room temperature, with the disorderly accumulation of more and more props on the table: "*By the end of the play the table has collected an inventory of objects*," a stage direction reads (ibid.: 19–20, original italics). However, performances of *Arcadia* also show that the possibilities of literature can suspend the strict rules of science: while the arrow of time is irreversible for Valentine and Hannah who can only glimpse the past in books and drawings, the plot strand set in the nineteenth century and the action set in the contemporary period are equally present for the audience. Even more, past and present coexist on stage, and in the last scene, they intertwine when characters from both time strands wear clothes from the early nineteenth century and appear on stage together: characters in both periods simultaneously grapple with the consequences of Thomasina's mathematical discovery and finally abandon the task to waltz. While the characters are oblivious to their counterparts and their parallel movements, the more distanced position of the audience allows them to see the similarities and appreciate this "island of order" in the fractal play. As mathematics appears as theme, metaphor, and structural device, it is fully integrated into the play, and grasping the detailed concepts is necessary to follow the plot. In this way, *Arcadia* presents math as an integral part of and means to understand life and literature.

CONCLUSION AND OUTLOOK: ARTIFICIAL INTELLIGENCE

Twentieth-century representations of mathematics explore it in terms of its traditional characteristics of certainty and exactitude, but also as a field of surprising uncertainty, freedom, and creative possibility that can seem to stand outside Enlightenment rationalization. The examples from poetry, drama, novelistic fiction, and the visual arts have shown that math has often been used to symbolize reason, whose excesses twentieth-century voices have lamented, and that math has frequently been seen as part of dystopian visions of inhuman order and control. At the same time, mathematics can exemplify truth that no perversion of power can touch, "prove" the inevitability of irrational and imaginary elements in life, and work as a metaphorical means to understand human life in its complexity. Whether based on historical mathematicians and concepts or using fictional characters and counterfactual entities, literary texts in the twentieth century have explored notions of mathematics as dependent on social and historical circumstances that influence the emergence and acceptance of discoveries. For example, the understanding of chaos theory changes with the availability of computers in *Arcadia*, and *Mason & Dixon* suggests that changing attitudes toward pre-Enlightenment and non-European worldviews led to the vanishing of the Vortex and the 365.25-degree circle. This historicity is also important for texts that establish certain concepts, or math in general, as specifically modern and as a model for twentieth-century art. Literary fiction in particular identifies similarities between its own imaginative worlds and the self-referential, self-consistent realm of mathematics. In the field of literature and science studies, which developed in the 1980s, mathematics initially played a very limited role. However, in the last decade a quickly growing subfield of literature and mathematics studies has emerged that examines how literature employs mathematical themes, metaphors, symbols, and structures to create and communicate images of math, its practitioners, and its audiences.[7]

In the twenty-first century, artificial intelligence has come into the spotlight as an important topic in representations of mathematics, both as a theme and as a means of representation. Artificial intelligences are based on algorithms, thus on mathematics; so if in Zamyatin's *We* humans live according to the law of the multiplication table, artificial intelligences could be said to exist according to the law of algorithms. In Ian McEwan's novel *Machines Like Me* (2019), the humanoid robot Adam describes himself as "a self, created out of mathematics, engineering, material science and all the rest" (McEwan 2019: 234). The suicide-prone artificial intelligences in the novel struggle to find reasons to continue existing, but Adam gives three arguments for his will to live: "Mathematics . . . poetry, and love for you" (ibid.: 279). Adam, having fallen in love with a human woman and created poetry, draws on practical experience to suggest that math, art, and emotion are equally good reasons to live. While the notion that artificial intelligences feel love is predominantly the stuff of fiction, the idea that mathematically created entities create poetry has already become a reality.

Already in 1973, computer scientists and linguists designed a program to write murder mystery stories, each 2100 words long and taking no more than nineteen seconds to create. The "novel writer," as the developers called the automated linguistic tool, was programmed in the language FORTRAN V on a Univac 1108 and devised various stories of murder and detection. To give an example:

> The day was Thursday. The weather was rainy. A small pub was on a corner. John Buxley was in the pub. John Buxley asked for whisky on the rocks. John got a drink from the barman. John talked with Hume near the bar. Hume sang the Beatles's song. John Buxley was drunk. James said that Marion committed adultery. James thought that James was drunk. James was depressed. James left the pub. (Klein et al. 1979: 402)

More recently, Botnik studios fed the seven *Harry Potter* books by J.K. Rowling into a predictive text keyboard and then worked with the algorithm to compose a new chapter of Harry Potter's adventures: "Harry Potter and the Portrait of what Looked like a Large Pile of Ash." The result of the algorithm-human-collaboration cannot fool a Potter fan, with giveaways including: "To Harry, Ron was a loud, slow, and soft bird"; "The floor of the castle seemed like a large pile of magic"; and "'I'm Harry Potter,' Harry began yelling. 'The dark arts better be worried, oh boy!'" (Botnik studios 2017). However, algorithms, and thus ultimately mathematics, are likely to play a larger role in the future, not necessarily in replacing human creativity and human authors, but in being involved in the creative process. Already, the potential for algorithms that are able to learn and evolve raise questions about art and creativity, original and copy, and practical concerns regarding the identity of the author of an algorithmically produced artwork: who should be designated the author of the new Harry Potter chapter and gain any proceeds? The programmer? The person working with the text prediction? J. K. Rowling on whose work the algorithm was based? Similar questions could be and are being asked about visual art produced by algorithms. So while at the beginning of the twentieth century, artists praised mathematical language as an example of precision and used its symbols and concepts to make texts and paintings appear strange and develop new ways of representation, in the future, writers, painters, and programming "mathemagicians" are likely to draw more often on mathematical algorithms to create new forms of art.

NOTES

Chapter 1

1. On the connection with higher mathematics, see Leo Corry, "Axiomatics between Hilbert and the New Math: Diverging Views on Mathematical Research and Their Consequences on Education," *International Journal for the History of Mathematics Education*, 2, 2 (2007): 21–37.
2. Hélène Gispert and Gert Schubring, "Societal, Structural, and Conceptual Changes in Mathematics Teaching: Reform Processes in France and Germany over the Twentieth Century and the International Dynamics," *Science in Context*, 24, 1 (2011): 73–106, at 86.
3. Edward L. Thorndike, "The Nature, Purposes and General Methods of Measurements of Educational Products," in Guy Montrose Whipple ed., *The Seventeenth Yearbook of the National Society for the Study of Education, Part II: The Measurement of Educational Products* (Bloomington, IN: Public School Publishing Company, 1918), 16–24, at 16.
4. Philip Alston [Special Rapporteur on extreme poverty and human rights], "Extreme Poverty and Human Rights," *United Nations Report*, A/74/493 (11 October 2019), available online: https://undocs.org/A/74/493, p. 2 (accessed April 16, 2021).

Chapter 3

1. Luzin's 1930 monograph on analytic sets had a large impact on the further development of descriptive set theory throughout the world and even impacted general set theory by its emphasis on the continuum hypothesis. In 1938, Kurt Gödel proved the consistency of the continuum hypothesis with the remaining axioms of set theory ("The Consistency of the Axiom of Choice and the Generalized Continuum-Hypothesis," *Proceedings of the National Academy of Sciences*, 24 (1838): 556–7). Hence, the continuum hypothesis could never be disproved within set theory.

A quarter-century later, Paul Cohen showed that it could not be proved either, which is to say, CH was independent of the other axioms. See Paul J. Cohen, "The Independence of the Continuum Hypothesis," *Proceedings of the National Academy of Sciences*, 50 (1963): 1143–8, and "The Independence of the Continuum Hypothesis. II," *Proceedings of the National Academy of Sciences*, 51 (1964): 105–10.

2. A transcript of the hearings was almost miraculously unearthed during the 1990s, despite the attempts of an unknown interested person to remove it from the archives of the Academy of Sciences. (One recalls the aphorism of the writer Mikhail Bulgakov: Manuscripts don't burn.) This transcript was published with commentary in a monograph edited by Sergei S. Demidov, Boris V. Lyovshin, and others as *The Case of Academician Nikolai Nikolaevich Luzin* (1999). An English translation by Roger Cooke, published by the American Mathematical Society, Providence, RI, appeared in 2016.
3. The Russian term for a Banach algebra—*normed ring* (*normirovannoe kol'tso*)—came from the title of Izrail' M. Gel'fand's paper "Normierte Ringe," published in German, but in the Soviet journal *Matematicheskii Sbornik*, 9, 51 (1941): 3–24. Thus, Gel'fand—who had no doubt taken to heart the lesson provided by the Luzin affair—had decided to forego the wider reception that would have been available in a German journal, but still wanted the paper to be read by Western mathematicians, most of whom could not read Russian.
4. Nicolas Bourbaki, "The Architecture of Mathematics," *American Mathematical Monthly*, 57 (1950): 221–32, at 228–9.
5. Although not a main issue of concern for all the members, discussions were held in the meetings of the group in the late 1940s about a possible formal concept that would provide the underlying support for the general informal approach, and from which all of the individual instances could be defined as special cases. Such discussions led to no significant results and, later on, the rise of the functorial language created a whole new situation in this regard. See Corry (2004), ch. 7.
6. Pontryagin published this result in English and in the American journal *Annals of Mathematics*, thus providing an example of the kind of publishing that Stalin disapproved of, the kind that lay at the heart of the case that would soon be brought against Luzin. Like all scholars, the Soviets wanted their works to be read by professionals capable of understanding them. Publishing in a Soviet journal and in Russian was a double obstacle to communication between Soviet and Western mathematicians.
7. Translated by Roger Cooke from Aleksandr D. Aleksandrov, "An Overview of Mathematics," in Aleksandr D. Aleksandrov, Andrei N. Kolmogorov, and Mikhail A. Lavrent'ev eds., *Mathematics, Its Content, Methods, and Meaning* (Moscow: USSR Academy of Sciences Press, Moscow, 1956), 5–79, at 77 (Russian). This passage was omitted from the English translation of the book, published by the AMS in 1963.
8. Jean Dieudonné, *Foundations of Modern Analysis* (New York: Academic Press, 1960), v–vi.
9. A similar interesting case to mention is Bourbaki's treatment of "flat modules" in the book on *Commutative Algebra*, whose chapters first appeared between 1961 and 1965. This concept is also best understood in homological and categorical terms, but since the treatise had not handled such ideas before 1980, Bourbaki surprisingly

deviated here from its self-imposed norm of relying only on previously handled material. They included a whole section formulated in terms of concepts and techniques, which they promised to elaborate in a forthcoming volume where functors and categories, and in particular, Abelian categories would be treated. They also made an exception by referring interested readers to books outside the treatise. The volume on categories, however, never appeared.

10. In the United States, the immigrants proved to be more productive in mathematical research than their native-born American contemporaries. See George J. Borjas and Kirk B. Doran, "The Collapse of the Soviet Union and the Productivity of American Mathematicians," *Quarterly Journal of Economics*, 127, 3 (August 2012): 1143–203.

Chapter 4

1. This definition is given by David Mumford, "Why I am a Platonist," *EMS Newsletter* (December 2008): 27. The literature is abundant, see, e.g., the polemical debate between mathematician Alain Connes and neuroscientist Jean-Pierre Changeux in (Connes and Changeux 1998).
2. "On the Method of Theoretical Physics," Spencer Lecture 1933, in Albert Einstein, *Ideas and Opinions* (New York: Crown Publishers, 1954), 270–6.
3. Paul Dirac, "The Evolution of the Physicist's Picture of Nature," *Scientific American*, 208: 5 (1963): 53.
4. Hans Hahn, *Empiricism, Logic, and Mathematics: Philosophical Papers* (Heidelberg: Springer, 1980), xii.
5. See especially Henri Poincaré, *Science and Hypothesis* (London: Bloomsbury Academic, 2017, French original 1902); and the studies (Gray 2013) and (Verhulst 2012).
6. See G. H. Hardy, *A Mathematician's Apology* (Cambridge: Cambridge University Press, 2012), or Nicholas Bourbaki, "The Architecture of Mathematics," *American Mathematical Monthly*, 57, 4 (1950): 221–32.
7. Hermann Weyl, "On the New Foundational Crisis in Mathematics," in Paolo Mancosu ed., *From Brouwer to Hilbert: The Debate on the Foundations of Mathematics in the1920s* (Oxford: Clarendon Press, 1998), 86–122, at 99.
8. David Hilbert, "On the Grounding of Mathematics: First Report," in Mancosu (1998), 198–214.
9. Herbert Mehrtens, *Moderne - Sprache - Mathematik* (Frankfurt: Suhrkamp, 1990); see also Gray (2008).
10. See Graham (1993). In genetics, there was the famous Lysenko affair, with classical genetics seen as an expression of capitalist ideas, and the inheritance of acquired characters being promoted for ideological reasons.
11. For a highly readable presentation of the three authors and their fate, see Graham and Kantor (2009). This focuses on the Name Worshipping movement (a heterodox religious current, considered heretical by the Orthodox Church) and its persecution by the Soviets.
12. Kolmogorov's *Foundations of Probability*, his best known work, was published in German (Springer, 1933), as was Aleksandrov and Hopf's *Topologie* (Springer, 1935). In the trial, Aleksandrov acknowledged that he himself had published more abroad than had Luzin.

13. Words of Kolmogorov in 1963, translated from the Spanish version by Carlos Sánchez Fernández and Concepción Valdés Castro, *Kolmogórov, el zar del azar* (Tres Cantos: Nivola, 2003).
14. Martin Heidegger, "Modern Science, Metaphysics, and Mathematics," in *Basic Writings* (London: Harper, 2008), 275.
15. Ian Stewart, "The Mathematical Equation that Caused the Banks to Crash," *The Guardian*, February 11, 2012.
16. See Grothendieck, Letter to the Swedish Academy of Sciences, published in *Le Monde* and translated in the *Mathematical Intelligencer*, 11, 1 (1989). On the pathological end of the spectrum, one finds mathematician Theodore Kaczynski (known as the Unabomber) and his noteworthy publication, "Industrial Society and Its Future," published in the *New York Times* and the *Washington Post*.
17. Quoted from BBC News, March 24, 2010: "Maths Genius Urged to Take Prize," available online: http://news.bbc.co.uk/2/hi/europe/8585407.stm (accessed April 8, 2022).
18. Oswald Veblen, *Bulletin of the American Mathematical Society*, 30 (1924): 289.
19. Quoted in *In the Matter of J. Robert Oppenheimer, USAEC Transcript of Hearing Before Personnel Security Board* (1954).
20. Fields medalist Vladimir Voevodsky launched a project for new foundations linked with computer verification, based not on set theory but on a form of type theory, HoTT. Bill Thurston was quite wary of Voevodsky's notion of proof as formal verification.
21. As a striking example of this, see William Thurston, *Three-Dimensional Geometry and Topology* (Princeton, NJ: Princeton University Press, 1997).

Chapter 5

1. This distinction between the approaches of Planck and Einstein was forcefully argued in Kuhn ([1978] 1987).
2. The origins of quantum physics are discussed in Darrigol (1993), Kragh (1999), and Duncan and Janssen (2019). The first book encompasses the development of quantum physics from 1900 to the late 1920s and emphasizes the use of analogies. The second deals with the history of physics in the twentieth century in general. The third traces the early contributions to the old quantum theory facing new historiographical challenges.
3. For the history of different interpretations of quantum mechanics, see Freire Jr. et al. (2022).
4. For more about multielectronic systems, see Jagdish Mehra and Helmut Rechenberg, *The Historical Development of Quantum Theory*, 6 vols (New York: Springer, 1982–2001). This study provides a comprehensive account about many developments of quantum mechanics from their origins to the 1940s.
5. On the rise of the standard model, see Lilian Hoddeson eds., *The Rise of the Standard Model: Particle Physics in the 1960s and 1970s* (Cambridge: Cambridge University Press, 2010).
6. For a comprehensive account, see Renn (1997).
7. See Thorne (1994) for an account of the striking implications of the general theory of relativity.

Chapter 6

1. May died unexpectedly in 1977, but his colleagues constructed this paper based on a recording of his talk at Los Alamos.
2. Kenneth O. May, "Historiography: A Perspective for Computer Scientists," in Nicholas Metropolis, Jack Howlett, and Gian-Carlo Rota eds., *A History of Computing in the Twentieth Century* (New York: Academic Press, 1980), 11–18, at 12.
3. See, for instance, Slava Gerovich, *From Newspeak to Cyberspeak: A History of Soviet Cybernetics* (Cambridge, MA: MIT Press, 2002), for the USSR or Pierre Mounier-Kuhn, *L'informatique en France de la seconde Guerre Mondiale au Plan Calcul. L'émergence d'une science* (Paris: Presses de l'Université Paris-Sorbonne for France, 2010).
4. M. Wilkes, "Programme Design for a High-Speed Automatic Calculating Machine," *Journal of Scientific Instruments*, 26 (1949): 217–20.
5. The ACM is a US-based and well-known academic society for computing and computer science. It awards on an annual basis the so-called Turing award. It is often referred to as the Nobel Prize in computing.
6. Richard Hamming, "Intellectual Implications of the Computer Revolution," *American Mathematical Monthly*, 70, 1 (1963): 4–11.
7. See: http://oeis.org/wiki/Works, citing OEIS for over 2,000 papers that reference the encyclopedia in their work.
8. Derrick H. Lehmer, "Computer Technology Applied to the Theory of Numbers," in William J. Leveque ed., *Studies in Number Theory*, MAA Studies in Mathematics, vol. 6 (New York: Prentice Hall, 1969), 117–51.
9. Douglas Hartree, *Calculating Instruments and Machines* (Urbana, IN: University of Illinois Press, 1949).
10. See, for instance, the work by the RAND Corporation, which published a table of a million random numbers generated with a roulette wheel.
11. Editors of DATA-LINK, "Letter to the Editor: What's in a Name," *Communications of the ACM*, 1, 4 (1958): 6.
12. David Nofre, Mark Priestly, and Gerard Alberts, "When Technology became Language," *Technology and Culture*, 55, 1 (2014): 40–75.
13. Nicholas Metropolis, Jack Howlett, and Gian-Carlo Rota eds., Preface to *A History of Computing in the Twentieth Century* (New York: Academic Press, 1980).
14. As reported in *Zentralblatt*, one of two basic abstracting and reviewing service in pure and applied mathematics (consulted on March 9, 2021).
15. "Digital Intuition," editorial, *Nature*, 529 (2016): 437.

Chapter 7

1. For research focusing on literary modernism and mathematics, see: Brits (2017); Engelhardt (2018); Rodal (forthcoming). Regarding the argument that mathematics undergoes a modernist transformation partly comparable to that of literature and the arts, see Gray (2008).
2. For research on literature, art, and the fourth dimension, see: Blacklock (2018); Henderson (1983); Hickman (2005); Whitworth (2001).

3. David Hilbert, "On the Infinite," in Paul Benacerraf and Hilary Putnam eds., *Philosophy of Mathematics: Selected Readings* (Oxford: Blackwell, [1925] 1964), 134–51, at 141.
4. The quotation appears in Richard Bentley, *The Works of Richard Bentley*, vol. 3, edited by Alexander Dyce (London: Macpherson, 1838).
5. [Bernard de Fontenelle] *Théorie des tourbillons cartésiens; avec des réflexions sur l'attraction* (Paris: Guerin, 1752).
6. Yves Gingras, "What did Mathematics do to Physics?" *History of Science*, 39 (2001): 383–416, at 385.
7. In addition to the works cited in notes 1 and 2, recent research includes: Apostolos Doxiadis and Barry Mazur eds., *Circles Disturbed: The Interplay of Mathematics and Narrative* (Princeton, NJ: Princeton University Press, 2012); Lynn Gamwell, *Mathematics + Art; A Cultural History* (Princeton, NJ, and Oxford: Princeton University Press, 2016); Matthew Handelman, *The Mathematical Imagination: On the Origins and Promise of Critical Theory* (New York: Fordham University Press, 2019); Andrea K. Henderson, *Algebraic Art: Mathematical Formalism and Victorian Culture* (Oxford: Oxford University Press, 2018); Arielle Saiber and Henry S. Turner, "Mathematics and the Imagination: A Brief Introduction," *Configurations: Special Issue: Mathematics and Imagination*, 17, 1–2 (2009): 1–18; Robert Tubbs, *Mathematics in Twentieth-Century Literature and Art: Content, Form, Meaning* (Baltimore, MD: Johns Hopkins University Press, 2014); Robert Tubbs, Alice Jenkins, and Nina Engelhardt, eds., *Palgrave Handbook of Literature and Mathematics* (London: Palgrave Macmillan, 2020).

BIBLIOGRAPHY

Agar, John (2003), *The Government Machine*, Cambridge, MA: MIT Press.

Archibald, Thomas, Della Dumbaugh, and Deborah Kent (2014), "U.S. Mathematicians in World War I," in David Aubin and Catherine Goldstein eds., *The War of Guns and Mathematics: Mathematical Practices and Communities in France and its Western Allies around World War I*, 229–71, History of Mathematics, #42, Providence, RI: American Mathematical Society.

Aronowitz, Robert (2015), *Risky Medicine: Our Quest to Cure Fear and Uncertainty*, Chicago, IL: University of Chicago Press.

Aspray, William (1990), *John von Neumann and the Origins of Modern Computing*, Cambridge, MA: MIT Press.

Atwood, Margaret ([1985] 2010), *The Handmaid's Tale*, London: Vintage.

Aubin, David (1997), "The Withering Immortality of Nicolas Bourbaki: A Cultural Connector at the Confluence of Mathematics, Structuralism, and the Oulipo in France," *Science in Context*, 10 (2): 297–342.

Aubin, David, and Amy Dahan Dalmedico (2002), "Writing the History of Dynamical Systems and Chaos: *Longue Durée* and Revolution, Disciplines and Cultures," *Historia Matematica*, 29 (3): 273–339.

Aubin, David, and Catherine Goldstein, eds. (2014), *The War of Guns and Mathematics: Mathematical Practices and Communities in France and its Western Allies around World War I*, History of Mathematics, #42, Providence, RI: American Mathematical Society.

Beck, Ulrich ([1986] 1992), *Risk Society: Towards a New Modernity*, trans. Mark Ritter, Los Angeles, CA: Sage.

Beery, Janet L., Sarah J. Greenwald, Jacqueline A. Jensen-Vallin, and Maura B. Mast, eds. (2017), *Women in Mathematics: Celebrating the Centennial of the Mathematical Association of America*, New York: Springer.

Benjamin, Ruha (2019), *Race after Technology: Abolitionist Tools for the New Jim Code*, Cambridge: Polity Press.

Blacklock, Mark (2018), *The Emergence of the Fourth Dimension: Higher Spatial Thinking in the Fin de Siècle*, Oxford: Oxford University Press.

Botnik Studios (2017), "Harry Potter and the Portrait of What Looked like a Large Pile of Ash." Available online: https://botnik.org/content/harry-potter.html (accessed March 2, 2019).

Bouk, Dan (2015), *How Our Days became Numbered: Risk and the Rise of the Statistical Individual*, Chicago, IL: University of Chicago Press.

Brits, Baylee (2017), *Literary Infinities: Number and Narrative in Modern Fiction*, London: Bloomsbury.

Broch, Hermann ([1950] 2000), *The Guiltless*, trans. Ralph Manheim, Evanston, IL: Marlboro/Northwestern.

Campbell-Kelly, Martin, Kelly Aspray, Nathan Ensmenger, and Jeffrey R. Yost (2014), *Computer: A History of the Information Machine*, 3rd edn, Boulder, CO: Westview Press.

Campbell-Kelly, Martin, Mary Croarken, Raymond Flood, and Eleanor Robson (2003), *The History of Mathematical Tables: From Sumer to Spreadsheets*, Oxford: Oxford University Press.

Carson, John (2007), *The Measure of Merit: Talents, Intelligence, and Inequality in the French and American Republics, 1750–1940*, Princeton, NJ: Princeton University Press.

Connes, Alain, and Jean-Pierre Changeux (1998), *Conversations on Mind, Matter and Mathematics*, Princeton, NJ: Princeton University Press.

Cook, Marianna (2009), *Mathematicians: An Outer View of the Inner World*, Princeton, NJ: Princeton University Press.

Corry, Leo (1997), "The Origins of Eternal Truth in Modern Mathematics: Hilbert to Bourbaki and Beyond," *Science in Context*, 10 (2): 253–296.

Corry, Leo (2004), *Modern Algebra and the Rise of Mathematical Structures*, 2nd edn, Basel: Birkhäuser.

Corry, Leo (2006), "Axiomatics, Empiricism, and Anschauung in Hilbert's Conception of Geometry: Between Arithmetic and General Relativity," in José Ferreirós and Jeremy Gray eds., *The Architecture of Modern Mathematics*, 155–76, Oxford: Oxford University Press.

Corry, Leo (2008a), "Number Crunching vs. Number Theory: Computers and FLT, from Kummer to SWAC (1850–1960), and Beyond," *Archive for History of Exact Sciences*, 62 (4): 393–455.

Corry, Leo (2008b), "Writing the Ultimate Mathematical Textbook: Nicolas Bourbaki's *Éléments de mathématique*," in Eleanor Robson and Jacqueline Stedall eds., *The Oxford Handbook of the History of Mathematics*, 565–87, Oxford: Oxford University Press.

Darrigol, Olivier (1993), *From c-Numbers to q-Numbers: The Classical Analogy in the History of Quantum Theory*, Berkeley, CA: University of California Press.

De Mol, Liesbeth (2019), "Turing Machines," in Edward N. Zalta ed., *The Stanford Encyclopedia of Philosophy*, Winter 2019 edition. Available online: https://plato.stanford.edu/archives/win2019/entries/turing-machine/ (accessed June 10, 2022).

De Mol, Liesbeth, and Maarten Bullynck (2018), "Making the History of Computing: The History of Computing in the History of Technology and the History of Mathematics," *Revue de Synthèse*, 139 (3/4): 361–80.

De Mol, Liesbeth, and Marie-José Durand-Richard (forthcoming), "Calculating Machines and Numerical Tables—A Reciprocal History," in Dominique Tournès ed., *The History of Mathematical Tables*. Berlin: Springer.

Dostoyevsky, Fyodor ([1864] 2003), *Notes from Underground* and *The Double*, trans. Jessie Coulson, London: Penguin.

Doxiadis, Apostolos, and Barry Mazur, eds. (2012), *Circles Disturbed: The Interplay of Mathematics and Narrative*, Princeton, NJ: Princeton University Press.
Duncan, Anthony, and Michel Janssen (2019), *Constructing Quantum Mechanics, vol. I: The Scaffold 1900–1923*, Oxford: Oxford University Press.
Duren, Peter, ed. (1988–9), *A Century of Mathematics in America*, 2 vols, Providence, RI: American Mathematical Society.
Eliot, T. S. (1927), "A Commentary," *The Monthly Criterion*, 6 (4): 289–91.
Eliot, T. S. ([1931] 2015), "Triumphal March," in Christopher Ricks and Jim McCue eds., *The Poems of T.S. Eliot*, vol. I, 131–2, London: Faber & Faber.
Engelhardt, Nina (2018), *Modernism, Fiction and Mathematics*, Edinburgh: Edinburgh University Press.
Ensmenger, Nathan (2016), "The Multiple Meanings of a Flowchart," *Information and Culture*, 51 (3): 321–51.
Epple, M. (2006), "Felix Hausdorff's Considered Empiricism," in José Ferreirós and Jeremy Gray eds., *The Architecture of Modern Mathematics*, 263–89, Oxford: Oxford University Press.
Ferreirós, José ([1999] 2007), *Labyrinth of Thought: A History of Set Theory and its Role in Modern Mathematics*, 2nd edn, Basel: Birkhäuser.
Ferreirós, José (2008), "The Crisis in the Foundations of Mathematics," in Timothy Gowers ed., *Princeton Companion to Mathematics*, 142–56, Princeton, NJ: Princeton University Press.
Ferreirós, José, and Jeremy Gray, eds. (2006), *The Architecture of Modern Mathematics*, Oxford: Oxford University Press.
Freire, Jr, Olival, Guido Bacciagaluppi, Olivier Darrigol, Thiago Hartz, Christian Joas, Alexei Kojevnikov, and Osvaldo Pessoa, Jr, eds. (2022), *The Oxford Handbook of the History of Quantum Interpretations*, Oxford: Oxford University Press.
Galison, Peter (1997), *Image and Logic: A Material Culture of Microphysics*, Chicago, IL: University of Chicago Press.
Gamwell, Lynn (2016), *Mathematics + Art: A Cultural History*, Princeton, NJ: Princeton University Press.
Gerovich, Slava (2002), *From Newspeak to Cyberspeak: A History of Soviet Cybernetics*, Cambridge, MA: MIT Press.
Ghosh, Arunabh (2020), *Making it Count: Statistics and Statecraft in the Early People's Republic of China*, Princeton, NJ: Princeton University Press.
Gigerenzer, Gerd, Zeno Swijtink, Theodore Porter, John Beatty, and Lorenz Krüger (1989), *The Empire of Chance: How Probability Changed Science and Everyday Life*, Cambridge: Cambridge University Press.
Ginsburg, Mirra, ed. and trans. ([1923] 1970), *A Soviet Heretic: Essays by Yevgeny Zamyatin*, Chicago, IL: University of Chicago Press.
Gitelman, Lisa, ed. (2013), *"Raw Data" is an Oxymoron*, Cambridge, MA: MIT Press.
Goodstein, Judith R. (2018), *Einstein's Italian Mathematicians: Ricci, Levi-Civita, and the Birth of General Relativity*, Providence, RI: American Mathematical Society.
Gould, Stephen Jay ([1981] 1996), *The Mismeasure of Man*, revised edition, New York: Norton.
Gowers, Timothy, ed. (2008), *Princeton Companion to Mathematics*, Princeton, NJ: Princeton University Press.
Graham, Loren (1993), *Science in Russia and the Soviet Union: A Short History*, Cambridge: Cambridge University Press.
Graham, Loren, and Jean-Michel Kantor (2009), *Naming Infinity*, Cambridge, MA: Harvard University Press.

Gray, Jeremy (2006), "Modern Mathematics as a Cultural Phenomenon," in José Ferreirós and Jeremy Gray eds., *The Architecture of Modern Mathematics*, 371–96, Oxford: Oxford University Press.

Gray, Jeremy (2008), *Plato's Ghost: The Modernist Transformation of Mathematics*, Princeton, NJ: Princeton University Press.

Gray, Jeremy (2013), *Henri Poincare: A Scientific Biography*, Princeton, NJ: Princeton University Press.

Green, Judy, and Jeanne LaDuke (2008), *Pioneering Women in American Mathematics: The Pre-1940 PhD's*, History of Mathematics, #34, Providence, RI: American Mathematical Society.

Green, Judy, and Jeanne LaDuke. 2016. "Supplementary Material for *Pioneering Women in American Mathematics: The Pre-1940 PhD's*." Last updated January 15, 2016. Available online: http://www.ams.org/publications/authors/books/postpub/hmath-34-PioneeringWomen.pdf (accessed October 20, 2019).

Grier, David Allen (2005), *When Computers were Human*, Princeton, NJ: Princeton University Press.

Guerraggio, Angelo, and Pietro Nastasi (2005), *Italian Mathematics between the Two World Wars*, Basel: Birkhäuser.

Handelman, Matthew (2019), *The Mathematical Imagination: On the Origins and Promise of Critical Theory*, New York: Fordham University Press.

Harris, Michael (2008), "'Why Mathematics?' You Might Ask," in Timothy Gowers ed., *The Princeton Companion to Mathematics*, 966–76, Princeton, NJ: Princeton University Press.

Harris, Michael (2015), *Mathematics without Apologies: Portrait of a Problematic Vocation*, Princeton, NJ: Princeton University Press.

Heims, Steve J. (1980), *John von Neumann and Norbert Wiener*, Cambridge, MA: MIT Press.

Henderson, Andrea K. (2018), *Algebraic Art: Mathematical Formalism and Victorian Culture*, Oxford: Oxford University Press.

Henderson, Linda Dalrymple (1983), *The Fourth Dimension and Non-Euclidean Geometry in Modern Art*, Princeton, NJ: Princeton University Press.

Herf, Jeffrey (1984), *Reactionary Modernism: Technology, Culture and Politics in Weimar and the Third Reich*, Cambridge: Cambridge University Press.

Hickman, Miranda B. (2005), *The Geometry of Modernism: The Vorticist Idiom in Lewis, Pound, H.D., and Yeats*, Austin, TX: University of Texas Press.

Hobsbawm, Eric (1995), *The Age of Extremes: 1914–1991*, London: Abacus.

Hoddeson, Lillian, Laurie Brown, Michael Riordan, and Max Dresden, eds. (2010), *The Rise of the Standard Model: Particle Physics in the 1960s and 1970s*, Cambridge: Cambridge University Press.

Hodges, Andrew ([1983] 2014), *Alan Turing: The Enigma*, Princeton, NJ: Princeton University Press.

Hofstadter, Douglas R. (1979), *Gödel, Escher, Bach: An Eternal Golden Braid*, New York: Basic Books.

Hollings, Christopher D. and Reinhard Siegmund-Schultze (2020), *Meeting under the Integral Sign? The Oslo Congress of Mathematicians on the Eve of the Second World War*, History of Mathematics, #44, Providence, RI: American Mathematical Society.

Hyman, Louis (2012), *Debtor Nation: The History of America in Red Ink*, Princeton, NJ: Princeton University Press.

Igo, Sarah (2007), *The Averaged American: Surveys, Citizens, and the Making of a Mass Public*, Cambridge, MA: Harvard University Press.

Jungnickel, Christa, and Russell McCormmach (1986, 1990), *Intellectual Mastery of Nature: Theoretical Physics from Ohm to Einstein*, 2 vols, Chicago, IL: University of Chicago Press.

Katz, Victor, ed. (2007), *The Mathematics of Egypt, Mesopotamia, China, India and Islam: A Sourcebook*, Princeton, NJ: Princeton University Press.

Kidwell, Peggy Aldrich, Amy Ackerberg-Hastings, and David Lindsay Roberts, eds. (2008), *Tools of American Mathematics Teaching, 1800–2000*, Washington, DC/ Baltimore, MD: Smithsonian Institution/Johns Hopkins University Press.

Klein, Sheldon, John F. Aeschlimann, David F. Balsiger, Steven L. Converse, Claudine Court, Mark Foster, Robin Lao, John D. Oakley, and Joel Smith (1979), "Automatic Novel Writing: A Status Report," in János S. Petöfi ed., *Text Processing—Textverarbeitung; Research in Text Theory—Untersuchungen zur Texttheorie*, vol. 3, 338–412, Berlin: de Gruyter.

Kragh, Helge S. (1999), *Quantum Generations: A History of Physics in the Twentieth Century*, Princeton, NJ: Princeton University Press.

Krömer, Ralf (2007), *Tool and Object: A History and Philosophy of Category Theory*, Science Networks, Historical Studies, #32, Basel: Birkhäuser.

Kuhn, Thomas S. ([1978] 1987), *Black-Body Theory and the Quantum Discontinuity, 1894–1912*, 2nd edn, Chicago, IL: University of Chicago Press.

Lave, Jean (1988), *Cognition in Practice: Mind, Mathematics, and Culture in Everyday Life*, Cambridge: Cambridge University Press.

Lawrence, D. H. ([1915] 2011), *The Rainbow*. London: Vintage.

Lehto, Olli (1998), *Mathematics without Borders: A History of the International Mathematical Union*, New York: Springer.

Lorenat, Jemma, John McCleary, Volker Remmert, David E. Rowe, and Marjorie Senechal, eds. (forthcoming), *Max Dehn: A Polyphonic Portrait*, Providence, RI: American Mathematical Society.

Lorentz, G. G. (2002), "Mathematics and Politics in the Soviet Union from 1928 to 1953," *Journal of Approximation Theory*, 116 (2): 169–223.

Lupton, Deborah (2016), *The Quantified Self*, Cambridge: Polity Press.

MacKenzie, Donald (2004), *Mechanizing Proof: Computing, Risk and Trust*, Cambridge, MA: MIT Press.

MacKenzie, Donald (2006), *An Engine, Not a Camera: How Financial Models Shape Markets*, Cambridge. MA: MIT Press.

Mahoney, Michael (2011), *Histories of Computing*, ed. Thomas Haigh, Cambridge, MA: MIT Press.

Marinetti, Filippo Tommasso ([1912] 2009), "Technical Manifesto of Futurist Literature," in Lawrence Rainey, Christine Poggi, and Laura Wittman eds., *Futurism: An Anthology*, 119–25, New Haven, CT: Yale University Press.

Marinetti, Filippo Tommasso ([1914] 2009), "Geometric and Mechanical Splendor and the Numerical Sensibility," in Lawrence Rainey, Christine Poggi, and Laura Wittman eds., *Futurism: An Anthology*, 175–80, New Haven, CT: Yale University Press.

Marinetti, Filippo Tommasso, with Marcello Puma and Pino Masnata ([1941] 2009), "Qualitative Imaginative Futurist Mathematics," in Lawrence Rainey, Christine Poggi, and Laura Wittman eds., *Futurism: An Anthology*, 298–301, New Haven, CT: Yale University Press.

Marks, Harry M. (1997), *The Progress of Experiment: Science and Therapeutic Reform in the United States, 1900–1990*, New York: Cambridge University Press.

Mashaal, Maurice (2006), *Bourbaki: A Secret Society of Mathematicians*, Providence, RI: American Mathematical Society.

McEwan, Ian (2019), *Machines Like Me*, London: Jonathan Cape.
Mehra, Jagdish and Helmut Rechenberg (1982–2001), *The Historical Development of Quantum Theory*, 6 vols, New York: Springer.
Menand, Louis (2021), *The Free World: Art and Thought in the Cold War*, New York: Picador.
Moon, Bob (1986), *The "New Maths" Curriculum Controversy*, London: Falmer Press.
Muhammed, Khalil Gibran (2010), *The Condemnation of Blackness: Race, Crime, and the Making of Modern Urban America*, Cambridge, MA: Harvard University Press.
Muller, Jerry Z. (2018), *The Tyranny of Metrics*, Princeton, NJ: Princeton University Press.
Musil, Robert ([1913] 1990), "The Mathematical Man," in *Precision and Soul: Essays and Addresses*, ed. and trans. Burton Pike and David S. Luft, 39–43, Chicago, IL: University of Chicago Press.
Neufeld, Michael J. (2007), *Von Braun: Dreamer of Space, Engineer of War*, New York: Knopf.
O'Neil, Cathy (2016), *Weapons of Math Destruction: How Big Data Increases Inequality and Threatens Democracy*, New York: Crown.
Oreskes, Naomi, and Eric M. Conway (2010), *Merchants of Doubt: How a Handful of Scientists Obscured the Truth on Issues from Tobacco Smoke to Global Warming*, New York: Bloomsbury.
Orwell, George ([1948] 1990), *Nineteen Eighty-Four*, London: Penguin.
Pais, Abraham (1982), *Subtle is the Lord: The Science and the Life of Albert Einstein*, Oxford: Oxford University Press.
Parshall, Karen Hunger (2022), *The New Era in American Mathematics, 1920–1950*, Princeton, NJ: Princeton University Press.
Parshall, Karen Hunger, and Adrian C. Rice, eds. (2002), *Mathematics Unbound: The Evolution of an International Mathematical Research Community, 1800–1945*, History of Mathematics, #23, Providence, RI/London: American Mathematical Society/London Mathematical Society.
Phillips, Christopher J. (2015), *The New Math: A Political History*, Chicago, IL: University of Chicago Press.
Pier, Jean-Paul, ed. (2000), *Development of Mathematics 1950–2000*, Basel: Birkhäuser.
Pike, Burton, and David S. Luft, ed. and trans. ([1913] 1990), *Precision and Soul: Essays and Addresses*, 39–43, Chicago, IL: University of Chicago Press.
Polya, George (1945), *How to Solve It*, Princeton, NJ: Princeton University Press.
Porter, Theodore M. (1986), *Trust in Numbers: The Pursuit of Objectivity in Science and Public Life*, Princeton, NJ: Princeton University Press.
Priestley, Mark (2011), *A Science of Operations: Machines, Logic and the Invention of Programming*, Berlin: Springer.
Pynchon, Thomas ([1973] 2000), *Gravity's Rainbow*, London: Vintage.
Pynchon, Thomas (1997), *Mason & Dixon*, New York: Picador.
Rainey, Lawrence, Christine Poggi, and Laura Wittman, eds. (2009), *Futurism: An Anthology*, New Haven, CT: Yale University Press.
Renn, Jürgen, ed. (1997), *The Genesis of General Relativity*, 4 vols, Dordrecht: Springer.
Ricks, Christopher, and Jim McCue, eds. (2015), *The Poems of T.S. Eliot*, vol. I, London: Faber & Faber.
Roberts, David Lindsay (2012), *American Mathematicians as Educators, 1893–1923: Historical Roots of the "Math Wars"*, Boston, MA: Docent Press.

Robson, Eleanor, and Jacqueline Stedall, eds. (2008), *Oxford Handbook of the History of Mathematics*, Oxford: Oxford University Press.
Rodal, Jocelyn (forthcoming), *Modernism's Mathematics: From Form to Formalism*.
Rowe, David E. (2018), *A Richer Picture of Mathematics: The Göttingen Tradition and Beyond*, New York: Springer.
Rowe, David E., and Mechthild Koreuber (2020), *Proving it Her Way: Emmy Noether, a Life in Mathematics*, Cham: Springer Nature.
Rowe, David E., and Robert Schulmann (2007), *Einstein on Politics: His Private Thoughts and Public Stands on Nationalism, Zionism, War, Peace, and the Bomb*, Princeton, NJ: Princeton University Press.
Schappacher, Norbert (2022), *Framing Global Mathematics: The International Mathematical Union between Theorems and Politics*, Cham: Springer Nature.
Schweber, Sylvan (1994), *QED and the Men Who Made It: Dyson, Feynman, Schwinger, and Tomonaga*, Princeton, NJ: Princeton University Press.
Segal, Sanford L. (2003), *Mathematicians under the Nazis*, Princeton, NJ: Princeton University Press.
Siegmund-Schultze, Reinhard (2001), *Rockefeller and the Internationalization of Mathematics between the Two World Wars: Documents and Studies for the Social History of Mathematics in the 20th Century*, Basel: Birkhäuser.
Siegmund-Schultze, Reinhard (2009), *Mathematicians Fleeing from Nazi Germany: Individual Fates and Global Impact*, Princeton, NJ: Princeton University Press.
Sklar, Jessica K., and Elizabeth S. Sklar, eds. (2012), *Mathematics in Popular Culture: Essays on Appearances in Film, Fiction, Games, Television and Other Media*, Jefferson, NC: McFarland.
Smith, Barbara Herrnstein, and Arkady Plotnitsky (1997), *Mathematics, Science, and Postclassical Theory*, Durham, NC: Duke University Press.
Stoppard, Tom ([1993] 2000), *Arcadia*, London: Faber & Faber.
Tedre, Matti (2014), *The Science of Computing: Shaping a Discipline*, Boca Raton, FL: CRC Press.
Thorne, Kip S. (1994), *Black Holes and Time Warps: Einstein's Outrageous Legacy*, New York: W. W. Norton.
Thurston, W. (1994), "On Proof and Progress in Mathematics," *Bulletin of the American Mathematical Society*, 30 (2): 161–77.
Tournès, Dominique, ed. (forthcoming), *The History of Mathematical Tables*, Berlin: Springer.
Tubbs, Robert (2014), *Mathematics in Twentieth-Century Literature and Art: Content, Form, Meaning*, Baltimore, MD: Johns Hopkins University Press.
Tubbs, Robert, Alice Jenkins, and Nina Engelhardt, eds. (2020), *Palgrave Handbook of Literature and Mathematics*, London: Palgrave Macmillan.
van Dalen, Dirk (2013), *L.E.J. Brouwer—Topologist, Intuitionist, Philosopher: How Mathematics is Rooted in Life*, Heidelberg: Springer.
Verhulst, Ferdinand (2012), *Henri Poincaré—Impatient Genius*, New York: Springer.
Vucinich, Alexander (2002), "Soviet Mathematics and Dialectics in the Post-Stalin Era: New Horizons," *Historia Mathematica*, 29 (1): 13–39.
Weil, André (1992), *The Apprenticeship of a Mathematician*, Boston, MA: Birkhäuser.
Whitworth, Michael (2001), *Einstein's Wake: Relativity, Metaphor, and Modernist Literature*, Oxford: Oxford University Press.
Yandell, Benjamin (2002), *The Honors Class: Hilbert's Problems and Their Solvers*, Wellesley, MA: A. K. Peters.
Zamyatin, Yevgeny ([1921] 1993), *We*, trans. Clarence Brown, London: Penguin.

Zamyatin, Yevgeny ([1923] 1970), "On Literature, Revolution, Entropy, and Other Matters," in *A Soviet Heretic: Essays by Yevgeny Zamyatin*, ed. and trans. Mirra Ginsburg, 107–12, Chicago, IL: University of Chicago Press.

Zdravkovska, Smilka, and Peter L. Duren, eds. (1993), *Golden Years of Moscow Mathematics*, History of Mathematics, #6, Providence, RI: American Mathematical Society.

INDEX

Locators in *italics* refer to illustrations.